SOIL FERTILITY

ENVIRONMENTAL SCIENCE, ENGINEERING AND TECHNOLOGY SERIES

The Amazon Gold Rush and Environmental Mercury Contamination
Daniel Marcos Bonotto and Ene Glória da Silveira
2009. ISBN: 978-1-60741-609-8

Environmental Effects of Off-Highway Vehicles
Douglas S. Ouren, Christopher Haas, Cynthia P. Melcher, Susan C. Stewart, Phadrea D. Ponds, Natalie R. Sexton Lucy Burris, Tammy Fancher and Zachary H. Bowen
2009. ISBN: 978-1-60692-936-0

Nitrous Oxide Emissions Research Progress
Adam I. Sheldon and Edward P. Barnhart (Editors)
2009. ISBN: 978-1-60692-267-5

Agricultural Runoff, Coastal Engineering and Flooding
Christopher A. Hudspeth and Timothy E. Reeve (Editors)
2009. ISBN: 978-1-60741-097-3

Fundamentals and Applications of Biosorption Isotherms, Kinetics and Thermodynamics
Yu Liu and Jianlong Wang (Editors)
2009. ISBN: 978-1-60741-169-7

Conservation of Natural Resources
Nikolas J. Kudrow (Editor)
2009. ISBN: 978-1-60741-178-9

Forest Canopies: Forest Production, Ecosystem Health and Climate Conditions
Jason D. Creighton and Paul J. Roney (Editor)
2009. ISBN: 978-1-60741-457-5

Directory of Conservation Funding Sources for Developing Countries: Conservation Biology, Education and Training, Fellowships and Scholarships
Alfred O. Owino and Joseph O. Oyugi
2009 ISBN: 978-1-60741-367-7

Process Engineering in Plant-Based Products
Hongzhang Chen
2009. ISBN: 978-1-60741-962-4

Soil Fertility
Derek P. Lucero and Joseph E. Boggs (Editors)
2009. ISBN: 978-1-60741-466-7

SOIL FERTILITY

DEREK P. LUCERO
AND
JOSEPH E. BOGGS
EDITORS

Nova Science Publishers, Inc.
New York

For permission to use material from this book please contact us:
Telephone 631-231-7269; Fax 631-231-8175
Web Site: http://www.novapublishers.com

LIBRARY OF CONGRESS CATALOGING-IN-PUBLICATION DATA

Soil fertility / [edited by} Derek P. Lucero and Joseph E. Boggs.
p. cm.
Includes index.
ISBN 978-1-60741-466-7 (hardcover)
1. Soil fertility. I. Lucero, Derek P. II. Boggs, Joseph E.
S596.7.S636 2009
631.4'22--dc22

2009018831

Published by Nova Science Publishers, Inc. ✢ New York

CONTENTS

PREFACE

Soil fertility refers to the amount of nutrients in the soil, which is sufficient to support plant life. Soil fertility is critical in sustainable farming and needs to be considered not only for crop productivity but also for the protection of aquatic environments. Fertile soil has an abundance of plant nutrients including nitrogen, phosphorus and potassium, an abundance of minerals as well as an abundance of organic matter. This book discusses new developments in the field.

Chapter 1 - It has been well established that naturally occurring beneficial soil microbes play a vital role in the maintenance of agricultural soil fertility to ensure food production for a still-growing world population. Beneficial soil microbes are involved in a wide variety of activities that benefit soil fertility, plant growth and agricultural sustainability. These soil microbes elicit their beneficial effect by fixing atmospheric nitrogen into bioavailable ammonia and nitrate, converting insoluble phosphate such as calcium acid phosphate into soluble forms, improving and maintaining soil fertility, thereby improving plant growth and productivity. In this chapter, authors will discuss several different aspects of nitrogen-fixing and phosphate-solubilizing bacteria, with an emphasis on their applications as biofertilizers.

Chapter 2 - Municipal solid wastes are an increasing problem worldwide. The most environmentally friendly way of disposing of the organic fraction is to compost it and use it as soil-improvement in agriculture, as it both disposes of the waste and increases soil fertility. An experiment has been running for 11 years outside Vienna comparing bio-waste compost with mineral fertilisation and no fertilisation. The site is characterised by high natural fertility and dry climate in the growing season. Several soil and crop parameters were measured during the experiment, including yield, mineral nitrogen and organic carbon and nitrogen. Simulation models provide an excellent tool to integrate such data and provide an overall understanding of the system. The authors used this data set to parameterize and compare two simulation models, EPIC and CENTURY. These models share essentially the same carbon turnover model, but differ in the detail they describe crop growth and water balance. The authors found that EPIC predicted crop yield acceptably, while CENTURY underestimated it. Both models could predict total soil carbon and nitrogen reasonably well, despite large differences in predicted plant production. Both models predicted larger peaks of mineral nitrogen than observed and CENTURY overestimated and EPIC underestimated it overall. Neither of the models could predict soil water content very accurately. The models failed to account adequately for the slow-release fertiliser effect of compost, possibly because of failure to accurately simulate effect of low water availability and litter quality on decomposition rate.

Chapter 3 - Soil fertility is an important factor in enhancing crop productivity but is not commonly included in studies analyzing productivity and efficiency. In this study, authors examine the influence of selected soil fertility parameters on modern rice productivity utilizing a stochastic production frontier approach on a survey data from 21 villages in three agro-ecological regions of Bangladesh. Detailed crop input-output data were collected from 380 paddy rice (Oryza sativa) farmers. The soil fertility status in each region was determined by analysis of soil organic carbon, available nitrogen, phosphorus and potassium concentration. Results reveal that in addition to key production inputs, soil fertility also significantly affects the parameters of the production function. Available soil potassium and soil nitrogen significantly increase rice productivity whereas available soil phosphorus has an opposite effect. The soil organic carbon content also has a desirable positive effect but the influence is not significant. The mean technical efficiency of these modern rice farmers is estimated at 73%, implying that 37% [(100-73)/73] of the production can be increased by eliminating technical inefficiency alone. The authors results indicate that in policy terms technological initiatives should be targeted at measures to identify areas of lower soil fertility so that inherent soil-based productivity restrictions can be minimized. In part this will be facilitated by the transfer of indigenous knowledge from farmers in higher productivity areas, thus increasing rice production. Also, investment in rural infrastructure is suggested to improve technical efficiency.

JEL classification: O33; Q18; C21

Chapter 4 - As part of a larger study, this chapter investigates the influence of different fertilization practices on microbial parameters involved in P turn-over of Mediterranean soils. More precisely, the aim of the present study was to compare the effects of both mineral and organic amendments on microbial patterns involved in P recycling of two Mediterranean soils. With this aim, the effects of both Na and K phosphate salts (supplied as NaH_2PO_4 and KH_2PO_4, respectively), as well as sewage sludge compost (SSC) on both acid (APH) and alkaline (BPH) phosphatases and phosphodiesterase (PDI) activities, BPH/APH ratio, densities of culturable heterotrophic bacteria and available P (P Olsen) were investigated in two Mediterranean vineyard soils originating from calcareous and siliceous mineral parent materials. All of the amendments were carried out in order to provide soil with a P total quantity equivalent to 0.5 g P_2O_5 per kg of soil, and variables were measured after 25 and 87 d of incubation under field conditions. Globally, results showed that neither phosphate salts nor SSC had significant effects on P mineralizing enzyme activities, nor on the densities of culturable heterotrophic bacteria. On the contrary, all of these amendments resulted in a significant increase in available P of the two soils considered. Results also showed a significant effect of the incubation time on several variables, especially APH and BPH activities, as well as available P. Indeed, between the two sampling dates most of these enzyme activities significantly decreased, whereas P availability was improved, SSC amended soils showing the more important variations in available P contents. This fertilization experiment also revealed that soil origin had only a significant effect on BPH activities, and that the nature of the cation (i.e., Na or K) within the different P salts used did not show any significant effect on the different variables investigated.

Chapter 5 - Lectins are proteins with multiple sites that bind specific sugars and therefore can agglutinate cells or other substrates with those sugars on the surface. This property is exploited in detecting human ABO blood type, defined based on the presence of distinct surface sugars that result in agglutination of blood cells in response to appropriate lectins. This capacity to bind to specific sites on cell surfaces is of interest for its potential in treating human diseases including cancer, AIDS, and diabetes. In plants, species-specific lectins are secreted from root tips into the extracellular environment of agronomically important crops including legumes and cereals. The function, stability and distribution of these proteins after export from root cells into the extracellular environment remain unexplored. Experiments to examine predictions of the hypothesis that root-secreted lectins influence soil structure and fertility, rhizosphere microbial communities, and root nutrient cycling are warranted.

Chapter 6 - Leaf-cutting ants modify soil fertility through two mechanisms. First, the building, enlargement, and maintenance of nests ants affect soil structure, porosity and density. Second, leafcutters collect and concentrate vegetal material inside their nests to maintain their fungus culture, the food for most of the colony. As a result of this process, ants generate a huge quantity of organic waste that is deposited in nest cavities or dumps on the soil surface. This organic waste is very rich in nutrients. The content of organic Carbon, Nitrogen, Potassium, Phosphorous, and Magnesium is, in average, between 20 and 50 times greater in refuse dumps than in adjacent, non-nest soils. The process of waste deposition, thus, notably increases the nutrient content of soils around nests. Consequently, plants inhabiting nest areas often show greater abundance, growth rate, foliar and root biomass, and reproduction rate than plants outside nest areas. This positive effect on plants might scale-up and affect the structure of vegetation assemblages and the balance between trees and herbs at landscape scale. Several factors affect the contribution of leaf-cutting ants to soil fertility. Particularly, extreme weather, low abundance of palatable vegetation for leafcutters, competition with introduced mammalian herbivores and frequent fires reduce ant foraging rates, nest abundance, and therefore their production of organic waste. Since leaf-cutting ant activity and nest density is strongly dependent on the availability of pioneer or ruderal plant species, the strength of their contribution to soil fertility could be more important in early successional environments and disturbed habitats. Ant-nest areas should be especially protected because they are hot spot of plant diversity and core of plant succession. From a restoration point of view, the nutrient-rich refuse dumps can be employed as natural, free-access and ecologically sustainable fertilizer to improve soil fertility in degraded areas.

Chapter 7 - Soil fertility is critical in sustainable farming and needs to be considered not only for crop productivity but also for the protection of aquatic environments. Evidence suggests that soil microbial biomass is a flexible nutrient store, and with appropriate management such a nutrient store could have potential dual benefits for improved crop productivity and enhanced nutrient retention in soils to minimize nutrient export from agricultural land to environmental waters. To date, most research has focused on stimulating the release of nutrients from the microbial biomass pool for crop uptake. By contrast, authors understanding of microbial biomass as a soil regulator to buffer nutrient supply and potential leaching loss is still limited. Such a role however could be usefully explored in heavily weathered and nutrient-deficient sandy soils under Mediterranean climate conditions, such as soils in the coastal region of southwestern Australia. Here authors present authors views on the critical aspects of microbial immobilisation in relation to soil fertility and nutrient retention, and propose a conceptual framework for such a role of microbial nutrient retention. This is supported by previously reported data on microbial immobilisation and nutrient retention. The critical issues such as the size of soil microbial biomass and the time scale for biomass turnover, as well as its potential mobility in soil matrices are discussed along with the research needs in understanding the role of soil microbial processes in sustained soil fertility and environmental conservation.

Chapter 8 - Several simulation models for carbon and nitrogen turnover exist (e.g Parton et al., 1987; Jenkinson et al., 1987; Williams, 1990; Hansen et al., 1993; Bradbury et al., 1993). Most of them are mainly for agricultural systems, but some can deal with grasslands and forests as well. The interest in model development has been driven first by an interest in predicting nitrogen leaching and later by an interest in green house gas emissions and soil carbon storage. Modelling nitrogen availability and leaching when large and known quantities of nitrogen fertilisers are added is relatively easy. It is much more difficult to predict availability of nitrogen and other nutrients under natural conditions and low input agriculture. The problem is that under such conditions, predicting nutrient levels at any one time depend on accurate prediction of mineralisation and immobilisation rates. These are temporally and spatially variable, and the factors that control them are only partially known (e.g. Vinten et al., 2002; Silva et al., 2005). Some measurable soil parameters are known to be related to soil fertility, such as carbon and clay content (Sylvia et al., 1998). However, soil fertility can not be fully predicted from measurable parameters, so the concept of soil fertility is not fully understood. Soil classification systems (for example FAO classification) usually attach a measure of suitability for agriculture to each soil type, but this is based on experience rather than an understanding of how the fertility comes about. Process based models can only describe processes that are understood, and therefore struggle to describe natural soil fertility adequately. An alternative approach is to introduce soil fertility as calibrated factor based on soil type in the models (Fisher et al., 2002).This can be a good way to get relatively accurate predictions of yield, but does not really improve authors understanding of the concept of soil fertility. Furthermore, when also other parameters such as green house gas emission need to be predicted, a more quantitative understanding of the concept of soil fertility is urgently needed. This will require both a better understanding of the soil itself and the interaction between soil and inputs of fertilisers and organic material.

Chapter 9 - Sustaining soil fertility is a major agricultural constraint in the Amazon Basin (Tiessen, et al., 1994), thus shifting cultivation accompanied with slash and burn agriculture is the prevailing agricultural practice in the humid tropics.

In addition to the predominant and unproductive Ferralsols and Acrilols, an exceptional dark soil is well known by the indigenous people and colonists for its sustained soil fertility. According to its dark color and origin ,the soil was termed *Terra Preta de Indio*. Smith (1879) and Katzer (1903) were among the first who described the *Terra Preta's* properties and presumed its cultural origin. Later their assumptions were strengthened by the soil scientists Sombroek (1966) and Zech, et al. (1990). *Terra Preta* contains significantly more carbon (C), nitrogen (N), calcium (Ca), and phosphorus (P), and the cation exchange capacity (CEC), pH value, and base saturation are significantly higher in *Terra Preta* soils than in the surrounding Ferralsols and Acrisols (Zech, et al., 1990; Glaser, et al., 2000). *Terra Preta* soils contain up to 70 times more black C than the adjacent soils. Due to its polycyclic aromatic structure, black C is chemically and microbially stable and persists in the environment over centuries (Glaser, et al., 2001b). Today, the anthropic origin of *Terra Preta de Indio* is generally accepted and its fertility is most likely linked to an accumulation of P and Ca, associated with bone apatite (Zech, et al., 1990; Lima, et al., 2002) and black C as charcoal (Glaser, et al., 2001a). The evidence that *Terra Preta* was manmade, and thus the proven feasibility to transform one of the most infertile soils into one of the most productive soil inspired charcoal research.

Slash and Char was described as an alternative to slash and burn (Steiner, 2007) and (Steiner, et al., 2004b) observed that charcoal is currently used by Amazonian settlers to improve soil fertility. If a forest is burned, only around 2-3% of the above-ground C is converted into charcoal (Fearnside, et al., 2001), but charcoal production can capture 50% of the above-ground C. If the charcoal is not used as fuel (e. g., soil amendment) it has a very high recalcitrance against biological or chemical decay and stores the carbon over centuries or millennia. *Slash and char* is an alternative agricultural method producing charcoal out of the aboveground biomass instead of converting it to CO_2 through burning. If re-growing resources are used, *slash and char* could establish as a significant carbon sink and could be an important step towards sustainability and SOM conservation.

On a global scale, the total carbon release flux due to fire is of the order of 4-7 Pg of carbon per year. This flux is almost as large as the rate of fossil fuel consumption (about 6 Pg per year in 1990) (Goudriaan, 1995). Fearnside (2000) calculated a total net emission of carbon from tropical land uses, equivalent to approximately 29% of the total anthropogenic emission from fossil fuels and land-use change. These numbers emphasize the potential for C management if only biomass is utilized being ablaze each year.

As a result of intensive research done in the 1980s, carbonized materials are formally authorized for use as soil amendment material in Japan. Thus Japan used 27% of the total charcoal consumption (50,835 Mg in 1999) for purposes other than fuel. By far the biggest proportion was used for agriculture (30.6%) followed by livestock industries (22.3%), gardening and golf courses (7.6 %) (Okimori, et al., 2003). A Japanese company established charcoal production at an Indonesian tree plantation for pulp production. Their feasibility study with conventional charcoal-making methods showed that 77,000 Mg charcoal could be produced per year, and the carbon emission reductions by the project reaches 62,000 Mg C yr^{-1} (= 230,000 Mg CO_2 yr^{-1}) at an annual wood harvest of 10,000 ha (Okimori, et al., 2003).

Chapter 10 - The Himalayan mountain range stretches from Afghanistan in the west to Myanmar in the East covering a distance of over 3500 km. Nepal is a densely populated, highly diverse and ecologically fragile country situated within the Central Himalaya between India and China. The inhabitants of the Himalayan region depend heavily on subsistence agriculture as the mainstay and primary source of livelihood. Natural resources in Nepal, in particular, have come under tremendous pressures due to the ever-increasing human and livestock populations and demand for food, fiber, construction material and energy. This has led to degradation of the resource base including, forests, water and arable land. Numerous workers have reported a steady decline in the fertility status of agricultural and forest soils of Nepal. Moreover, high rates of soil erosion and nutrient loss have been reported from degraded forest land, grazing lands and sloping agricultural fields. Other factors contributing to the fertility and productivity decline of soils include: increasing trend of chemical fertilizer use, decreasing amounts and quality of farmyard manure/compost applied, and increasing cropping intensification. Sustainable options for maintaining and enhancing soil fertility and productivity must address increasing the return of crop residues and other organic materials to soil, reducing dependence on inorganic fertilizers, and shifting to alternative fertilizer sources (such as bio-fertilizers) and cropping systems (such as agroforestry and permaculture).

In: Soil Fertility
Editors: D. P. Lucero and J. E. Boggs
ISBN: 978-1-60741-466-7

Chapter 1

SOIL FERTILITY ENHANCING MICROORGANISMS AND THEIR APPLICATION AS BIOFERTILIZERS

***Lily Lili Liu**[*1] **and Shengwu Ma**[2]*
[1] Chemistry and Life Sciences College, Tianjin Normal University, Tianjin 300074, PR China
[2] Department of Biology, University of Western Ontario, London, ON, Canada N6A 5B7

ABSTRACT

It has been well established that naturally occurring beneficial soil microbes play a vital role in the maintenance of agricultural soil fertility to ensure food production for a still-growing world population. Beneficial soil microbes are involved in a wide variety of activities that benefit soil fertility, plant growth and agricultural sustainability. These soil microbes elicit their beneficial effect by fixing atmospheric nitrogen into bioavailable ammonia and nitrate, converting insoluble phosphate such as calcium acid phosphate into soluble forms, improving and maintaining soil fertility, thereby improving plant growth and productivity. In this chapter, we will discuss several different aspects of nitrogen-fixing and phosphate-solubilizing bacteria, with an emphasis on their applications as biofertilizers.

1. INTRODUCTION

With the global population rapidly growing, the world demand for food places an immense pressure on agricultural lands. According to predictions by the United Nations [1], by 2025, an increase of 40-45% in food production could be required to sustain the global population. To meet this challenging demand, modern agricultural practice has been relying more heavily on the use of chemical fertilizers to increase crop productivity. For example,

* Corresponding author. E-mail address: liulilisd@nankai.edu.cn; Tel. 86-22-13821449397; Fax: 86-222-350-8800.

global chemical fertilizer use was merely 27 million tons in 1959 and 1960; it increased five times to 141 million metric tons over a forty-year period ending in 2000. Furthermore, the projected fertilizer demand for the year 2020 is 220 million metric tons [2, 3]. While the use of chemical fertilizers has an immediate effect on the increase of food productivity, there have been increased concerns over the potential harmful environmental and human health effects associated with long-time overuse of chemical fertilizers [2, 3]. Excessive use of chemical fertilizers depletes the overall health of soil, which reduces a plant's ability to absorb nutrients and eventually results in a reduction of crop yield. As a consequence, more fertilizer is required to achieve the same crop yield. As large quantities of chemical fertilizer are not absorbed by plants and are lost to the environment, the use of chemical fertilizers can cause surface and groundwater pollution, algae bloom and oxygen depletion [4, 5]. Excessive use of chemical fertilizers also has negative impacts on air quality via ammonia emissions which can come down with rainfall or as direct deposition, acidifying the soil and water, and can even contribute to greenhouse gas emissions [4, 5]. Since a healthy soil is the key to sustainable agriculture production, use of eco-friendly biofertilizers is therefore highly desirable and may provide a valuable solution.

Biofertilizers, commonly known as microbial fertilizers or microbial inoculants, are artificially multipled cultures of certain soil microorganisms which can improve the fertility of the soil by providing essential and renewable sources of nutrients such as nitrogen and phosphoras to plants. Biofertilizers are natural, environment friendly, non-toxic, low-cost and safe to use [6,7]. The main sources of biofertilizers are from bacteria, fungi, and cyanobacteria (blue-green algae). Living cells of nitrogen-fixing or phosphate-solubilizing bacteria represent the most important and frequently used biofertilizer inoculants. Mycorrhizal fungi comprise a group of organisms that colonize roots of the majority of land plant species on earth and are known to confer an array of benefits to plants. They can absorb phosphorus from the soil and pass it onto the plant. Moreover, mycorrhizal fungi can promote water absorption by plants even under water-stressed conditions. Mycorrhizal fungi may also produce plant growth promoting hormones and other chemicals to enhance the tolerance of plants to high soil temperatures or help plants to resist various soil- and root-borne pathogens, and even toxic heavy metals [8]. The inoculation of mycorrhizal fungi has been shown to benefit a variety of crop plants, with a grain yield increase on average of 15 to 30% under field conditions [9]. Cyanobacteria are a morphologically diverse group of Gram-negative photosynthetic bacteria formerly known as blue-green algae. This group of bacteria includes unicellular forms, as well as filamentous forms that contain only vegetative cells or cells differentiated into nitrogen-fixing heterocysts [10]. The ability to photosynthesize and fix nitrogen, together with great adaptability to various soil types, makes cyanobacteria widely distributed in nature, although occupying mostly marine and freshwater habitats. Due to its nitrogen-fixing characteristics, cyanobacteria have been investigated as bioinoculants in agriculture to enhance crop productivity. The effect of adding dried cyanobacteria to soil on rice yield was first studied in the 1950s in Japan, with an average increase in rice productivity in the range of 15-20% [11]. The effect of cyanobacterial inoculants on other crops such as wheat has also been demonstrated [12].

In this chapter we provide an overview of soil fertility enhancing microorganisms. We will particularly focus on nitrogen-fixing and phosphate-solubilizing bacteria, as they represent two of the most important groups of microorganisms that play a pivotal role in maintaining natural soil fertility. We discuss their application and effectiveness as microbial biofertilizers in soil fertility and crop production.

2. Nitrogen-Fixing Bacteria and Their Application as Biofertilizers

Nitrogen (N_2) is essential for plant growth and it is a component of all proteins, enzymes and metabolic processes involved in the synthesis and transfer of energy. Insufficient quantity of N_2 limits plant production in both natural and agricultural ecosystems [13]. The atmosphere contains approximately 80 percent N_2 ($3.8X10^{18}$ kg), implying that above every hectare of ground there is ~80,000 tonnes of N_2, but unfortunately, it is unavailable to plants [13]. Gaseous N_2 has to be "fixed" into ammonia (NH4) or other forms such as nitrate (NO3) before it can be absorbed and used by most plants. N_2 fixation can occur as a result of natural non-biological processes such as lighting strikes, however, the amount of N_2 fixed via these non-biological processes accounts for only a small percentage, approximately 10% of the total fixed N_2. Roughly 90% of the fixed N_2 is accomplished through natural biological processes [14]. It is estimated that biological N_2 fixation on a global scale may reach a value of 175 million metric tons of fixed N_2 per year [15], representing a critical and key process in sustainable land management. N_2-fixing bacteria are a key player in biological N_2 fixation. In fact, bacteria are the only organisms capable of taking gaseous N_2 and combining it with hydrogen to make ammonia [15]. N_2-fixing bacteria are generally endemic to most soil types, but in the natural state, they only comprise a relatively small percentage of the total microbial population. N_2-fixing bacteria contain nitrogenase, an oxygen-sensitive enzyme that catalyzes chemical reactions to break the N_2 triple bond. Biological N_2 fixation can be represented by the following equation, in which two moles of ammonia are produced from 1 mole of N_2 gas, at the expense of 16 moles of ATP and a supply of electrons and protons (hydrogen ions):

$$N_2 + 8H^+ + 8e^- + 16\ ATP = 2NH_3 + H_2 + 16ADP + 16\ Pi$$

Nitrogenase requires the products of about 20 genes for its synthesis and activity [16]. Nitrogenase activity is deactivated in the presence of oxygen and therefore all N_2-fixing bacteria must operate under oxygen-free (anaerobic) conditions. All of the N_2-fixing organisms including cyanobacteria are prokaryotes. They can be free living or form symbiotic associations with plants or with other organisms (e.g. protozoa).

2.1. Symbiotic Nitrogen Fixers

There are three groups of commonly recognized symbiotic N_2-fixing microorganisms: rhizobia, actinomycetes (Frankia) and cyanobacteria. Filamentous actinobacteria from the genera *Frankia* form actinorhizal root nodules on diverse woody trees and shrubs, and constitute the major N_2-fixing symbioses in temperate forests, contributing significant amounts of available N_2 to forest land ecosystems [17]. However, the symbiotic relationship of *Frankia* with its woody trees is still not well understood. Cyanobacteria of the genus

Nostoc can form N_2 fixing symbiotic associations with a diverse variety of plants [18]. Under the influence of the plant partner, N_2-fixing cyanobacteria undergo cellular differentiation into motile filaments called hormogonia and into heterocysts. The hormogonia serve as the infective units in the establishment of an association, whereas heterocysts, specialized N_2-fixing cells, are the site of N_2 fixation in these cyanobacteria [18]. From a crop productivity point of view, rhizobial bacteria is seen as the most important N_2-fixing organism. Rhizobium invades roots of legumes, which are of great agricultural importance as a food crop, triggering the formation of a new organ, the root nodule, in which N_2-fixation occurs [19]. The *Rhizobium*-legume symbiosis represents one of the most efficient N_2-fixing systems known to date. In this unique association, plants provide a source of energy and an ecological niche for the bacterium, which in return synthesizes ammonia for the host plant. At the global scale, the *Rhizobium*-legume symbiosis provides a quantity of fixed nitrogen equivalent to that produced by the entire chemical fertilizer industry [20], and thus plays a major ecological and economic role. The *Rhizobium*-legume symbioses is the primary source of fixed nitrogen in land-based systems, providing well over half of the biological source of fixed nitrogen [20]. Moreover, N_2 fixed symbiotically by the *Rhizobium*-legume association represents a renewable source of N_2 for agriculture [21]. N_2 fixation rates of various legume crops and pasture species are commonly found in the range of 200 to 300 kg of N ha^{-1} $year^{-1}$ [21]. Yield increases of crops planted after harvesting of legumes are often equivalent to those expected from application of 30 to 80 kg of chemical N_2 fertilizer ha^{-1}.

Rhizobia are Gram-negative soil bacteria and they can be easily isolated from soils where leguminous plants are grown. The primary distinguishing characteristic of *Rhizobium* species is their ability to nodulate leguminous plants. *Rhizobia* are predominantly aerobic chemoorganotrophs, grow well in the presence of oxygen and can utilize a wide range of relatively simple carbohydrates and amino compounds. *Rhizobia* can also live independently in the soil, but they cannot fix nitrogen when living freely. Free living rhizobia have a different morphology from the bacteria found in root nodules. In general, cells of free living *Rhizobia* are regular in structure, appearing as straight rods, whereas cells of the *Rhizobium* bacteria living symbiotically inside the root nodules are differentiated into bacteroids having irregular shapes. The *Rhizobium* appears to be host specific, and each species can only nodulate a limited number of plant hosts, e.g. the pea is the host plant to *Rhizobium leguminosarum* biovar *viciae*, whereas the clover acts as a host to *R. leguminosarum* biovar *trifolii* and alfalfa serves as the host plant for *R. melilota.* Based on growth rate, symbiotic host ranges and other physiological properties, they are currently split into six genera: *Rhizobium, Sinorhizobium, Mesorhizobium, Bradyrhizobium, Azorhizobium* and *Allorhizobium* [22, 23]. Numerous studies have revealed that the formation of the nodules on leguminous hosts require cross-talk between plant and bacterial partners and this interaction determines host specificity. Plants secrete isoflavonoids that induce the bio-synthesis of bacterial lipo-chitooligosaccharides which are called Nod factors [24]. This rhizobial nodulation signal (i.e. Nod factor), initiates a cascade of events leading to the nodule primodium formation followed by entry of the bacteria into the nodule primodium via root hair cells [25]. Once the bacteria inhabit the nodule, the bacterium and the plant cooperate in fixation and assimilation of atmospheric N_2. Rhizobial *nod* genes have been found to reside on its symbiotic plasmid. Many of them have now been cloned and characterized. Based on their different biological functions, the *nod* genes can be subdivided into common *nod* genes and the special *nod* genes that are specifically involved in the host-specific interaction between rhizobia and their leguminous plant hosts. The common *nod* genes, such as *nodABC,* are present in all *Rhizobium* species, and are functionally interchangeable between species. Mutations in these genes cause complete nodulation failure. Common *nod* genes are involved in cell division and root hair deformation. In contrast to the common *nod* genes, host specific *nod* genes are not conserved because alleles from various rhizobial species cannot substitute for one another on different hosts. The main roles of host specific *nod* genes are to determine which *Rhizobium* strain infects which legumes [26].

2.2. Non-Symbiotic Nitrogen Fixers

Non-symbiotic N_2-fixing bacteria comprise a highly diversified group of organisms. They can be classified into at least 25 different genera. The most common and important genera include *Azotobacter, Azospirillum, Acetobacter, Azorhizobium, Azoarcus, Burkholderia, Clostridium, Enterobacteriaceae and Herbaspirillum* [27]. They occur in different habitats and can be aerobic, anaerobic or facultative anaerobic. For example, *Azotobacter* and *Azospirrillum* live in aerobic conditions, whereas *Clostridium* are obligate anaerobes. Some species of *Clostridium* are sensitive to oxygen concentrations as low as 0.5%. The sensitivity to molecular oxygen generally restricts *Clostridium* spp. to anaerobic areas such as water, submerged soil, rumina, and intestines [27].

N_2 fixation by non-symbiotic bacteria can also be a significant source of N_2 in cropping systems. In laboratory conditions, the species of *Azotobacter* are known to fix an average 10 mg of N/g of sugar in pure culture on a nitrogen free medium. However, the efficacy of N_2 fixation by non-symbiotic bacteria in natural conditions can be critically affected by the availability of carbon, energy sources in the soil and environmental conditions (soil moisture and temperature). On the other hand, non-symbiotic N_2-fixing bacteria have evolved many different ways of obtaining energy. For example, certain bacteria colonize the rhizosphere and the root surface, using root exudates as an energy source. These bacteria are often referred to as "associative N_2-fixing bacteria". Associative N_2-fixing bacteria are mostly represented by strains of *Azotobacter, Azospirillum* and *Acetobacter.* The phenominon of associative symbiosis was first described by Dobereiner and colleagues in 1972 [28]. They showed that *Azotobacter paspali* grows as a sheath around the root of the sand grass *Paspalum notatum* but with no development of nodules. Both the plant and the bacterium benefit greatly from this associative relationship, as the survival of *A. paspali* away from the plant is poor. The other bacteria can live inside the plant tissue without causing any symptoms of plant disease. They usually live in the intracellular spaces in the plant vascular system, take dissolved N_2 gas from the sap flow and convert it into ammonium for use by plants. As a reward, plants provide the bacteria with products of photosynthesis as a source of energy. These bacteria are often referred to as "endophytic N_2-fixing bacteria". Endophytic N_2-fixing bacteria include common strains of *Herbaspirillum, Acetobacter* and *Azospirillum* [29]. Since endophytic N_2-fixing bacteria live inside the plant, this renders them less vulnerable to competition from other soil-borne bacteria and are shielded from various biotic and abiotic stresses. An increasing number of endophytic N_2-fixing bacteria has now been isolated and characterized, and many of them were found to contain high N_2 fixation activity [29]. The N_2 fixation ability of non-symbiotic N_2-fixing bacteria is also dependent on the presence of the enzyme nitrogenese. Unlike the *Rhizobium*-legume symbiosis during which the N_2 fixed in root nodules is directly supplied to the plant, only a small proportion of N_2 fixed by non-symbiotic bacteria may become directly available to the plant. However, further benefits from non-symbiotic fixation may arise on a larger time scale due to long-term liberation of N_2 sequestered in microbial biomass by microfaunal grazers, a process with outstanding importance in the rhizosphere [29, 30].

2.3. Application of Nitrogen Fixing Bacteria as Biofertilizer

The utilization of N_2-fixing microbes for formulating biofertilizers for legumes as well as nonlegumes has had a long history. Biofertilizers have now been increasingly used to improve soil fertility and crop productivity in many countries, especially in the developing countries such as China, India, Cuba, Thailand, Vietnam, and the Philippines. In some provinces in China, the use of biofertilizers has even become mandatory. In general, rhizobial biofertilizer is far more developed than other types of biofertilizers, and its utilization can be traced back more than 100 years ago when two German scientists, Nobbe and Hiltner, demonstrated the advantage of mixing rhizobial bacteria with the seed at planting. The first patent regarding biofertilizer was filed in 1896 and it was related to inoculation of the soil with *Rhizobium* for cultivating leguminous plants [31].

Following the discovery of rhizobial biofertilizer, free living non-symbiotic N_2-fixing bacteria such as *Azotobacter* were also investigated as a potential biofertilizer for non-leguminous crops such as wheat, barley, maize, oat, sugarcane, sunflower, mustard, sesamum, linseed, tea and coffee, as well as many types of forest, fruit and flower plants [27, 32, 33, 34, 35]. Inoculation of soil or seed with *Azotobacter* inoculants has been shown to be effective in increasing yields of crops in soil with poor organic matter content [34, 35]. It is now clear that in addition to provide plants with fixed N_2, many strains of *Azotobacter* produce plant growth promoting substances such as indole acetic acid (IAA), gibberllins and vitamin-B, as well as antifungal substances. *Azotobacter* also causes the mineralization of fixed phosphate in soil, consequently increasing the uptake of phosphate in plants and enhancing stress resistance in plants [35].

Azospirillum was also evaluated as potential biofertilizers for agriculture. Soil application or seed inoculation of *A. lipoferum* resulted in a 22% increase in rice grain yield in field experiments [36], or enhanced phosphate and ammonia uptake by the plants [37]. A 30% yield increase was even reported for wheat inoculated with *A. brasilense* [38]. The yield increases in plants inoculated with *Azospirillum* inoculants have been attributed to increased availability of fixed N_2 together with some indirect effects exerted by *Azospirillum*. *Azospirillum* inoculation has been shown to enhance both root and root hair growth, resulting in significant increase of N_2 and mineral uptake. It has also been reported that plants treated with *Azospirillum* produce antifugal and antibacterial compounds as well as growth regulators at higher levels [35, 38].

3. Phosphate Solubilizing Bacteria

In addition to N_2, phosphorus (P) is another key element in plant growth and productivity. P plays an indispensable role in photosynthesis, respiration, energy storage and transfer, cell division and enlargement, protein and nucleic acid synthesis in the living plant. It is vital to seed formation and root development as well [39]. P is widely distributed in soil in both organic and inorganic forms. However, the majority of it is unavailable for plants to uptake due to its high reactivity with calcium (Ca), iron (Fe) or aluminium (Al), which leads to P precipitation. Inorganic P in acidic soils is associated with Al and Fe compounds, whereas calcium phosphates are the predominant form of inorganic P in calcareous soils. In general, the total P in soil accounts approximately for 0.04-0.10%, only 1.00-2.50% of which exists in a soluble ionic form (Pi) that can be absorbed by plants [40]. It was estimated that 74% of cultivated land in China is deficient in P. The P deficiency in soils has become a limiting factor for plant growth and productivity, especially after plants have gained sufficient N_2 nutrition [41, 42]. To circumvent the problem of P deficiency, the addition of P fertilizers has become a common practice in modern agriculture. However, a large portion of soluble inorganic P applied to soil as a chemical fertilizer is rapidly immobilized soon after application and becomes unavailable to plants [43], creating a demand for suitable alternatives to mobilize the fixed form of P. Soil microorganisms are able to mobilize insoluble mineral P in a more environmentally friendly and sustainable manner.

The involvement of microorganisms in the solubilization of inorganic P was known as early as 1903 [7]. Liu et al. [44] isolated a strain of phosphobacteria, designated as 9320-SD, from chestnut type soil in China and studied its *in vitro* P-solubilizing activity using both precipitated P agar plate and liquid broth culture methods. They demonstrated that 9320-SD can grow on an agar medium with insoluble powdered tricalcium phosphate ($Ca_3(PO_4)_2$) as the only P source, as evidenced by the production of a visible clear zone around each individual bacterial colony on the plate (See Figure 1). The formation of a clear/halo zone on the plate is due to the excretion of organic acids into the surrounding medium. To measure the rate of P solubilization, cells of 9320-SD were inoculated into liquid broth containing $Ca_3(PO_4)_2$ as the only P source, and incubated at 37^0C with shaking. They found that the amount of soluble P in the liquid medium was as high as 47% of the original weight of $Ca_3(PO_4)_2$ after four days of incubation. Liu et al. [45] additionally studied the kinetics of transforming $Ca_3(PO_4)_2$ into plant-available P by 9320-SD, and revealed that the extent of soluble P was positively correlated with the concentration of 9320-SD at logarithmic growth phase. Their relationship can be expressed by the following equation:

$Y = 4.54 \lg x + 11.23$ (Y: quantity of plant available phosphorus; X: bacterial concentration).

Using a scanning electron microscope (SEM), Liu et al. [6] also studied the different stages of P solubilization by 9320-SD. Samples of powdered $Ca_3(PO_4)_2$ were taken from 9320-SD cultured liquid medium or 9320-SD minus control liquid medium at different incubation periods. SEM images of the surface of the powered mineral P suspended in cultured 9320-SD liquid medium showed the erosion of the P particles, whereas no erosion was seen on the surface of P particles from control medium (see Figure 2). These results indicate that 9320-SD has strong P solubilizing ability and may have the potential for use as an effective P biofertilizer.

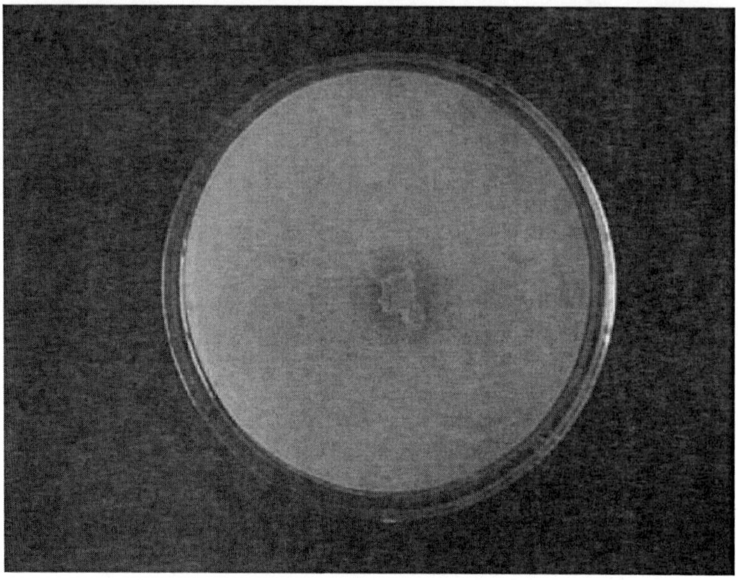

Figure 1. Phosphate solubilization by phosphobacterium 9320-SD, as indicated by hallow formation around the growth on agar plate with 1% insoluble-phosphate tricalcium phosphate ($CaHPO4$) as the only P source.

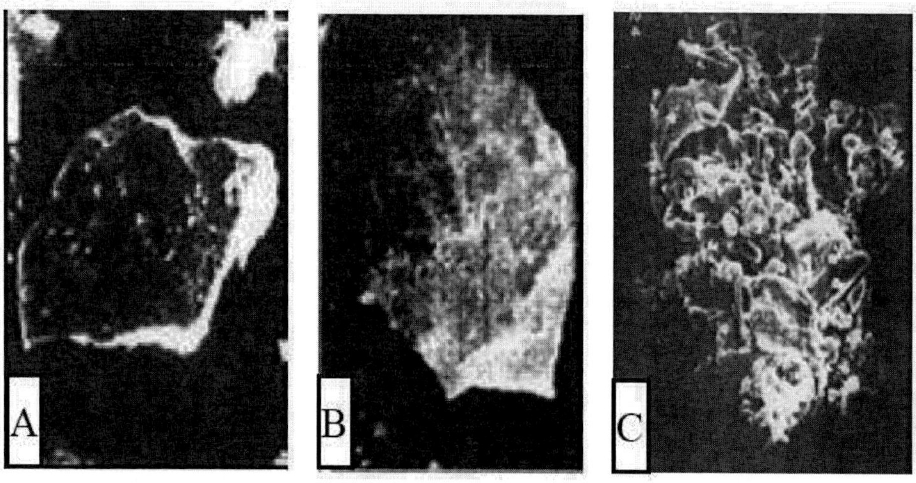

Figure 2. SEM (scanning electron microscopy) images of the surface of the powdered mineral phosphates suspended in liquid medium cultured with or without phosphobacterium 9320-SD. (a) The normal surface of powdered mineral phosphates; (b) The surface of the powdered mineral phosphates from liquid medium without the bacterial inoculation; (c) The surface of the powdered mineral phosphates from liquid medium after 168 h of inoculation with phosphobacterium 9320-SD.

It is now well established that P-solubilizing microorganisms are abundant in the soil. It is estimated that P-solubilizing microorganisms may constitute 20 to 40% of the culturable population of soil microorganisms. A significant proportion of these microorganisms can be isolated from rhizosphere soil [46, 47]. Ponmurugan and Gopi [48] studied the distribution pattern and population density of P-solubilizing bacteria in cultivated soils and showed that the population density of P-solubilizing bacteria is the highest in the rhizosphere soils of the groundnut ($14.9X10^5$/g soil dry wt.) followed by cotton, maize and the least in the soils of sorghum. The variation in the population of P-solubilizing bacteria is likely attributed to many soil factors such as soil nutrients, pH, moisture contents, organic matter and some soil enzyme activities.

Taxonamical studies of P-solubilizing microorganisms have suggested that they belong to a highly diversified group of organisms including bacteria, actinomycetes and fungi. Even within the bacterial group, they consists of taxonomically different and phylogenetic largely unrelated genera such as *Bacillus, Erwinia, Agrobacterium, Serratia, Flavobacterium, Enterbacter, Micrococcus, Azotobacter, Bradyrhizobium, Salmonella, Chromobacterium, Alcaligenes, Arthrobacter, Burkholderia, Thiobacillus,* and *Escherichia* [46, 47]. Among the P-solubilizing bacteria examined, species from *Bacillus, Pseudomonas,* and *Enterobacter* were shown to have higher P solubilizing activity [47]. The precise mechanism for P solubilization by different P-solubilizing microorganimsms remains mostly unknown. It has been suggested that organic acids produced by P-solubilizing organisms, which causes the soil pH to drop, is the major driving force for mobilization of mineral phosphates [47, 49]. Such organic acids are usually of low-molecular weight and may include acetate, lactate, oxalate, tartarate, succinate, citrate, gluconate, ketogluconate and glycolate [47, 49, 50]. These acids are able to chelate the cations bound to P through their hydroxyl and carboxyl groups, thereby converting it into soluble forms [51]. Bolan et al. [52] reported that the addition of organic acids to the soils with a high P-fixing capacity decreased the P sorption on the clay surfaces, favored the solubilization of rock phosphate, and increased dry matter of ryegrass (*Lolium rigidium*) and plant P uptake. This has so far become the most convincing evidence to support the role of oraganic acids in solubilizing mineral P. However, it should be realized that P solubilization is a complex phenomenon which depends on many factors such as nutritional, physiological and growth conditions of the P-solubilizing organism [53, 54]. Indeed, Goldstein [55] showed that P solubilization involves the direct oxidation pathway to produce gluconic acid and 2 keto-gluconic acid by the bacterial enzyme glucose dehydrogenase. Gluconic acid can solubilize insoluble P to release soluble P through the oxidation of glucose.

3.1. Application of P Solubilizing Bacteria as Biofertilizer

The use of P biofertilizers in the form of microbes to increase the availability of soluble P in soils for plant growth has attracted increasing attention due to its potential advantages, such as the reduced dependence on the utilization of chemical P fertilizer and increased crop yield and quality [7, 56]. Chemical P fertilizers are primarily manufactured from phosphate rocks that according to their origin, may contain various toxic heavy metals (As, Al, Cd, Pb and Hg), radionuclides and fluorides. The long-continued application of P fertilizers can redistribute and elevate heavy metal and fluorine concentrations in soil profiles. Consequently, their availability for plants can transfer these toxic heavy metals to the food chain, threatening human health and well-being. Moreover, chemical P fertilizers are costly. In addition to mobilizing P for plant use, P-solubilizing microbes exhibit other traits beneficial to plants, such as production of phytohormones, antibiotics, siderophores, vitamins, antifungal substances and hydrogen cyanide [47, 57]. Several reports have also suggested that

P-solubilizing microbes increase the efficiency of biological nitrogen fixation, and enhance the availability of trace elements such as Fe, Zn, *etc.* [7, 47, 58].

There have been an increased number of reports on plant growth promotion by P-solubilizing bacteria after their inoculation in soil or plant seeds. Rodriguez and Fraga [47] showed the yield increase of tomatoes, onions, potatos, bananas, citrics, and coffee in field conditions as a result of inoculation with a strain of *Burkholderia cepacia,* which is currently used as a commercial biofertilizer in Cuba. Lifshitz et al. [59] were the first to report that inoculation with *Pseudomonas putida* stimulated growth of roots and shoots, and increased P uptake in canola. Hameeda et al. [60] tested the biofertilizer effect of *Serratia marcescens* strain EB 67 and *Pseudomonas* sp. CDB 35 on maize under both glasshouse and field conditions. They showed a 99% increase with EB 67 and a 94% increase with CDB 35 in plant biomass (dry weight) under glasshouse conditions. Under field conditions, there was a 66% and 50% increase in plant biomass with EB 67 and a 51% and 18% increase with CDB 35, when measured at the respective 48 and 96 days after sowing. Furthermore, the grain yield of field-grown maize treated with EB 67 or CDB 35 had a 85% or 65% increase compared to the uninoculated control. They also showed that the inoculated strains of EB 67 and CDB 35 could still be traced back from the rhizosphere of maize even 96 days after sowing [60].

Our laboratory [6] studied the biofertilizer effect of phosphobacterium 9320-SD on winter wheat under both pot and field conditions. To produce a biofertilizer, the cell suspension culture of 9320-SD was mixed with peat as a carrier and the final density of viable cells was approximately 10^9 CFU (colony-forming unit) per gram of mixture. For pot trials, pots were filled with unsterial soil supplemented with the biofertilizer and seeds of winter wheat were planted. For field trials, the experiment was conducted in 18 randomly-chosen plots (each measured 3m x 3m) that had six treatment and three replates per treatment. Five treatments contained the biofertilizer applied at 300, 600, 900, 1,200, and 1,500 kg/km^2, respectively. Control plots were treated with 1,500 kg/km^2 of autoclaved 9320-SD biofertilizer. The winter wheat was planted and the biofertilizer was applied as the basal fertilizer in the seed row. Although 9320-SD biofertilizer showed little effect on plant height in both pot and field trials, they showed a significant increase in both the fresh and dry biomass compared to controls. Furthermore, the total P content in leaf sheath of the treated pot and field grown plants was found to be significantly higher than that of control fertilizer treated plants. These results suggest that the 9320-SD biofertilizer can offer phosphorus nutrition that promotes the growth of winter wheat in both pot and field conditions.

The use of mixed cultures or co-inoculation of P-solubilizing bacteria with other rhizospheric microorganisms may have an even better effect on plant growth and yield. Zaidi and Khan [61] demonstrated a synergistic effect of combined inoculation of the P-solubilizing bacterium *Pseudomonas striata* and the asymbiotic N_2-fixing bacterium *Azotobacter chroococcum* on the growth and yield of wheat as compared to their individual inoculations. Greater effectiveness of the combined inoculation was attributed to a better use of poorly soluble P sources (rock P). Similarly, combined inoculation of the P-solubilizing bacterium *Agrobacterium radiobacter* and the asymbiotic N_2-fixing *Azospirillum lipoferum* was shown to result in a greater increase in grain yield of barley as compared to single inoculations in both pot and field conditions [62]. Wani et al. [63] reported that combined use of the symbiotic N_2-fixing *Mesorhizobium* or the asymbiotic N_2-fixing *Azotobacter* and the P-solubilizing bacterium *Bacillus* or *Pseudomonas* had a much better stimulatory effect on the growth, yield, grain protein and uptake of N_2 and P of field-grown chickpea. It is now believed that compared to single inoculation, mixed inoculants could provide the plants with more balanced nutrition or offer the plants with conditions more suitable for N_2 fixation and/or easier to obtain plant-available soil P or some trace elements such as Fe and Zn [47].

4. CONCLUSION

Nitrogen and phosphorus are known to be essential nutrients for plant growth and development. Large quantities of chemical fertilizers are tranditionally used to replenish soil N and P for the attainment of high crop yields, resulting in high costs, severe environmental contamination and degradation of natural resources. There has currently been a growing interest in sustainable agriculture to alleviate detrimental effects associated with the overuse of chemical fertilizers by intensive agriculture practices. N_2-fixing and P-solubilizing bacteria may be important to plant nutrition by increasing N and P uptake by the plants, and play a significant role in the biofertilization of crops that would reduce the need for chemical fertilizers and decrease adverse environmental effects. Results from both greenhouse and field experiments have supported the ability of these bacteria to act as bioinoculants to increase plant growth and yield. However, more investigations will be required before the widespread adoption and use of biofertilizers. As the agronomic effectiveness of biofertilizers can be greatly affected by various environmental and biological factors such as soil conditions (pH, organic matter content, moisture level, availability of N and P, and indogenous soil microorganisms), selection of suitable bacteria as inoculants is important to the success of a biofertilizer product. The bacteria that are either best suited to the agro-eco system or to the target crop or are seen as highly likely to produce maximal beneficial effects should be selected for testing for their potential as biofertilizers. When *Rhizobium* strains are to be developed as biofertilizer inoculants, the selected bacteria should be matched to both the soil conditions and the specific legume being introduced to the soil to increase their survival in soil and competitiveness for nodulation, since rhizobial bacteria are host-specific. This may sometimes require additional work, such as screening new *Rhizobium* strains with higher competitive ability. Fortunately, there are a variety of techniques available for isolation of *Rhizobium* from either root nodules or soil samples. Similarly, many techniques are made available for the isolation, identification and screening of P-solubilizing bacteria. Yi et al. [64] recently reported the isolation of new P-solubilizing bacteria isolates from soils and their characterization. They showed that the isolates that produce exopolysaccharide (EPS) render the bacteria a greater P-solubilizing activity than the isolates producing no EPS. Their work encourages the search for newer strains or isolates with higher P-solubilizing or N_2-fixing activity that could be used as more effective biofertilizers. In the field application of any microbial inoculum, it is often essential to verify that the inoculated microorganisms cause the plant to respnd to them. Various strategies have now been developed to distinguish the inoculated strain from other endogenous strains. The use of marker genes such as antibiotic resistance genes or genes encoding firefly luciferase (luc) or bacterial luciferase (lux) to label the microbial inoculum is one of the strategies commonly used.

It has been established that naturalised or endogenous N_2-fixing or P-solubilizing bacteria are often more competitive under soil conditions than any strain used as an inoculum. It is therefore important to enhance the ability of inoculum strains to compete with native soil bacteria for root nodulation or for their thriving in the rhizosphere. One strategy is to develop and use recombinant bacterial strains as inoculants. Iniguez et al. [65] reported the construction of a broad host range plasmid (pHUTFXPAR) carrying genes that could enhance nodulation competitiveness through trifolitoxin production, and improve the efficacy of nitrogen fixation by oxidation of the H_2 evolved from nitrogenase. When plasmid pHUTFXPAR was transferred into *R. leguminosarum* bv. Phaseoli 127K105a and the resultant transformants were tested as an bioinoculum for bean plants (*Phaseolus vulgaris*) under field conditions, they found that there was a 12 to 14% yield increase in plants inoculated with the bacterial strain 127K105a(pHUTFXPAR) compared to plants either uninoculated or inoculated with the wild-type bacterium 127K105a (without plasmid pHUTFXPAR). Rodríguez et al. [66] reported the construction of recombinant P-solubilizing bacterial strains of *Burkholderia cepacia* IS-16 and *Pseudomonas* sp. containing plasmid DNA encoding *Erwinia herbicola* pyrroloquinoline quinone (PQQ) synthase responsible for mineral P solubilization, and demonstrated that the engineered strains produced higher clearing halos in plates with insoluble phosphorus as the unique P source, as compared to parentel strains.

In conclusion, biofertilizers offer a more economical and eco-friendly option for fertilization, and their potential benefits to the sustainable agriculture and our environment should never be overestimated.

ACKNOWLEDGMENTS

This work was supported by research grants from the National Natural Science Foundation of China (# 30470017) and Tianjin Municipal City's Priority Areas for Science & Technology Development (# 003122111-4).

REFERENCES

[1] Ongley ED. Introduction to Agricultural Water Pollution. In: Control of Water Pollution from Agriculture. Food and Agriculture Organization of the United Nations, Rome, Italy; 1996. p. 1-16.

[2] Pimentel D. Agricultural Chemicals and the Environment. In Lal R, Hansen DO, Uphoff N, Slack S, editors. Food Security and Environmental Quality in the Developing World, *Boca Raton, FL: CRC/Lewis Publishers;* 2002. p. 205–213.

[3] Jauhar PP, Khush GS. Importance of Biotechnology in Global Food Security. In: Lal R, Hansen DO, Uphoff N, Slack S, editors. Food Security and Environmental Quality in the Developing World, *Boca Raton, FL: CRC/Lewis Publishers;* 2002. p. 105– 126.

[4] Holford ICR. Soil phosphorus: its measurement, and its uptake by plants. *Aust. J. Soil Res.* 1997; 35: 227–239.

[5] Lo´pez-Bucio J, de la Vega OM, Guevara-Garcı´a A, Herrera-Estrella L. Enhanced phosphorus uptake in transgenic tobacco plants that overproduce citrate. *Nat. Biotech.* 2000; 18: 450–453.

[6] Chen Z, Ma SW, Liu LL. Studies on phosphorus solubilizing activity of a strain of phosphobacteria isolated from chestnut type soil in China. *Bioresour. Technol.* 2008; 99:6702-6707.

[7] Kucey RMN, Janzen HH, Leggett ME. Microbially mediated increases in plant-available phosphoras. *Ad. Agron.* 1989; 42:199-228.

[8] Young CC. Effects of phosphorus-solubilizing bacteria and vesicular-arbuscular mycorrhizal fungi on the growth of tree species in subtropical-tropical soils. *Soil Sci. Plant Nutr.* 1990; 36: 225-231.

[9] Solaiman MZ, Hirata H. Effect of arbuscular mycorrhizal fungi inoculation of rice seedlings at the nursery stage upon performance in the paddy field and greenhouse. *Plant and Soil.* 1997; 191:1-12.

[10] Castenholz RW, Waterbury JB. Oxygenic photosynthetic bacteria. Group I. Cyanobacteria. In: Staley JT, Bryant MP, Pfenning N, Holt JG, editors. Bergey's Manual of Systematic Bacteriology. *Williams and Wilkins, Baltimore* 1989; vol. 3, p. 1710-1789.

[11] Mishra U, Pabbi S. Cyanobacteria: A potential biofertilizer for rice. *Resonance J. Sci. Education.* 2004; 9:6-10.

[12] Karthikeyan N, Prasanna R, Nain L, Kaushik BD. Evaluating the potential of plant growth promoting cyanobacteria as inoculants for wheat. *Eur. J. Soil. Biol.* 2007; 43:23-30.

[13] Frink CR, Waggoner PE, Ausubel JH. Nitrogen fertilizer: retrospect and prospect. *Proc. Natl. Acad. Sci. U.S.A.* 1999; 96: 1175-1180.

[14] Graham PH, Vance CP. Nitrogen fixation in perspective: an overview of research and extension need. *Field Crop. Res.* 2000; 65:93-106.

[15] Berkum P, van Bohlool BB. Evaluation of nitrogen fixation by bacteria in association with roots of tropical grasses. *Microbiol. Rev.* 1980; 44:491-517.

[16] Lee S, Meletzus D, Sevilla M, Kennedy C. Characterization of a major nif/fix gene cluster in Acetobacter diazotrophicus, an endophyte of sugarcane. *J. Bacteriol.* 2000; 182:7088-7091.

[17] Schwintzer RC, Tjepkema DJ, editors. *The* Biology of Actinorhizal Plants. *Academic Press, San Diego, USA;* 1990.

[18] Meeks JC. Symbiosis between nitrogen-fixing cyanobacteria and plants. *Biosci.* 1998;48:266–276.

[19] Zahran HH. Rhizobium-legume symbiosis and nitrogen fixation under severe conditions and in an arid climate. *Micro Mol. Biology Rev.* 1999; 63:968-989.

[20] Brockwell J, Bottomley PJ, Thies JE. Manipulation of rhizobia microflora for improving crop productivity and soil fertility: a critical assessment. *Plant and Soil.* 1995; 174: 143–180.

[21] Peoples MB, Ladha JK, Herridge DF. Enhancing legume N_2 fixation through plant and soil management. *Plant Soil.* 1995; 174:83-101.

[22] Sadowsky MJ, Graham PH. Soil biology of the *Rhizobiaceae.* In: Spaink HP, Kondorosi A, Hooykaas PJJ, editors. The *Rhizobiaceae. Ameterdam:Kluwer;* 1998. p.155–172.

[23] Vance CP. 1997. Enhanced agricultural sustainability through biological nitrogen fixation. In: Legocki A, Bothe H, Puhler A, editors. Biological Fixation of Nitrogen for Ecology and Sustainable Agriculture. *Berlin: Springer-Verlag*; 1997. p. 179–186.

[24] Geurts R, Bisseling T. *Rhizobium* Nod Factor Perception and Signalling. *Plant Cell.* 2002; 14: s239–s249.

[25] Gage D J, Margolin W. Hanging by a thread: invasion of legume plants by rhizobia. *Curr. Opin. Microbiol.* 2000; 3:613–617.

[26] Long SR. Rhizobium symbiosis: nod factors in perspective. *Plant cell.* 1996; 8: 1885–1898.

[27] Tilak KVBR, Ranganayaki N, Pal KK, De R, Saxena AK, Shekhar N, et al. Diversity of plant growth and soil health supporting bacteria. Special Section: Microbial diversity. *Curr. Sci.* 2005; 89:136-150.

[28] Döbereiner J, Day JM, Dart PJ. Nitrogenase activity and oxygen sensitiviry of tbe *Paspaium notatum-Azofobacter puspali* association. *J. gen. Microbiol.* 1972; 71: 103-116.

[29] Reinhold-Hurek B, Hurek T. Life in grasses: diazotrophic endophytes. *Trends Microbiol.* 1998; 6:139-144.

[30] Egener T, Reinhold-Hurek B. Endophytic expression of *nif* genes of *Azoarcus* sp. BH72 in rice roots. *Mol. Plant-Microbe Interact.* 2000; 12:813-819.

[31] Nobbe F, Hiltner L. U.S. Patent 570 813. Inoculation of the soil for cultivating leguminous plants. 1896.

[32] Tilaki KVBR, Ranganayaki N, Pal KK, De R, Saxena AK, Nautiyal CS, et al. Diversity of plant growth and soil health supporting bacteria. *Curr. Sci.* 1995; 89:136-150.

[33] Yasari E, Patwardhan AM. Effects of *(Azotobacter* and *Azosprillium*) inoculants and chemical fertilizers on growth and productivity of canola (*Brassica napus* L.) effects. *Asian J. plant Sci.* 2007; 6: 77-82.

[34] Kader MA, Mian MH, Hoque MS. Effects of Azotobacter inoculant on the yield and nitrogen uptake by wheat. *Online J. Bio. Sci.* 2002; 2: 259-261.

[35] Kennedy IR, Choudhury ATMA, Kecskes ML. Non-symbiotic bacterial diazotrophs in crop-farming systems: can their potential for plant growth promotion be better exploited. *Soil Bio. Biochem.* 2004; 36:1229-1244.

[36] Balandreau J. The spermosphere model to select for plant growth promoting rhizobacteria. In: Kennedy IR, Choudhury ATMA, editors. Biofertilisers in Action. Rural Industries Research and Development Corporation. Canberra. 2002. p. 55–63.

[37] Murty MG, Ladha JK. Influence of Azospirillum inoculation on the mineral uptake and growth of rice under hydroponic conditions. *Plant and Soil.* 1988; 108, 281– 285.

[38] Okon Y, Labandera-Gonzalez CA. Agronomic applications of Azospirillum: an evaluation of 20 years world-wide field inoculation. *Soil Bio. Biochem.* 1994; 26: 1591–1601.

[39] Raghothama KG. Phosphate acquisition. *Annu. Rev. Plant Physiol. Plant Mol. Biol.* 1999; 50: 665-693.

[40] Lin CG. Agricultural chemistry of soil. In: Lin CG, editor. The Holes, Structures and Cultivable Peculiarly of Field Soil. *Beijing: Agriculture Press; 1990.* p. 56– 65.

[41] Borch K, Bouma TJ, Lynch JP, Brown KM. Ethylene: a regulator of root architectural responses to soil phosphorus availability. *Plant Cell Environ.* 1999; 22: 425–431.

[42] Kanako O, Motohiko K, Ho A, Ken-ichi K. Phosphorus application affects root length distribution and water uptake of upland rice in a column experiment. *Soil Sci. Plant Nutr.* 2004; 50: 257–261.

[43] Richa G, Khosla B, Reddy MS. Improvement of maize plant growth by phosphate solubilizing fungi in rock phosphate amended soils. *World J. Agric Sci.* 2007; 3: 481-484.

[44] Liu LL, Li SG. The screening and application of special bacteria for PK biofertilizer. *Universitatis Nankaiensis.* 1994; 27: 28-32.

[45] Zhang F, Zhu G, Liu LL, Chen Z. Kinetic properties of dissolving phosphorus by the phosphobacteria 9320-SD. *Electronic J. Biol.* 2007; 3: 69-73.

[46] Kucey RMN. Phosphate solubilizing bacteria and fungi in various cultivated and virgin Alberta soils. *Can. J. Soil Sci.* 1983; 63 :671-678.

[47] Rodriguez H, Fraga R. Phosphate solubilizing bacteria and their role in plant growth promotion. *Biotechnol. adv.* 1999; 17 :319-339.

[48] Ponmurugan P and Gopi C. Distribution pattern and screening of phosphate solubilizing bacteria isolated from different food and forage crops. *J. Agron.* 2006; 5: 600-604.

[49] Rodríguez H, Fraga R, Gonzalez T, Bashan Y. Genetics of phosphate solubilization and its potential applications for improving plant growth- promoting bacteria. *Plant Soil.* 2006; 287:15–21.

[50] Friedrich S, Platonova NP, Karavaiko GI, Stichel E, Glombitza F. Chemical and microbiological solubilization of silicates. *Acta Biotechnol.* 1991; 11:187–196.

[51] Kpomblekou K, Tabatabai MA. Effect of organic acids and release of phosphorus from phosphate rocks. *Soil. Sci.* 1994; 158: 442–452.

[52] Bolan N S, Naidu R, Mahimairaja S, Baskaran S. Influence of low-molecular- weight organic acids on the solubilization of phosphates. *Biology and Fertility of Soils.* 1994; 18:311-319.

[53] Reyes I, Bernier L, Simard RR, Antoun H. Effect of nitrogen source on the solubilization of different inorganic phosphates by an isoalte of *Penicillium rugulosum* and two UV-induced mutants. *FEMS Microbiol. Ecol.* 1999a; 28: 281- 290.

[54] Reyes I, Bernier L, Simard RR, Tanguay P, Antoun H. Characteristics of phosphate solubilization by an isolate of a tropical *Penicillium rugulosum* and two UV-induced mutants. *FEMS Microbiol. Ecol.*1999b 28: 291-295.

[55] Goldstein AH. Involvement of the quinoprotein glucose dehydrogenase in the solubilization of exogenous phosphates by Gram negative bacteria. In: Torriani- Gorini A, Yagil E, Silver S, editors. Phosphate in Microorganisms: Cellular and Molecular Biology. *ASM Press, Washington, DC;* 1994.p. 197–203.

[56] Goldstein AH. Bacterial solubilization of mineral phosphates: Historical perspective and future prospects. *Am. J. Alt. Agric.* 1986; 1:51-57.

[57] Kloepper JW, Lifshitz R, Zablotowicz RM. Free-living bacterial inocula for enhancing crop productivity. *Trends in Biotech.* 1989; **7:** 39-43.

[58] Gyaneshwar P, Kumar GN, Parekh LJ, Poole PS. Role of soil microorganisms in improving P nutrition of plants. *Plant Soil.* 2002; 245: 83-93.

[59] Lifshitz R, Kloepper JW, Kozlowshi M, Simonson C, Carlson J, Tipping M, et al. Growth promotion of canola (rapeseed) seedlings by a strain of Pseudomonas putida under gnotobiotic conditions. *Can. J. Microbio.* 1987; 33: 390-395.

[60] Hameeda B, Harini G, Rupela, OP, Wani SP, Reddy G. Growth promotion of maize by phosphate-solubilizing bacteria isolated from composts and macrofauna *Microbiol. Res.* 2008; 163:234-242.

[61] Zaidi A, Khan MS. Interactive effect of rhizospheric microorganisms on growth, yield and nutrient uptake of wheat. *J. Plant. Nutr.* 2005; 28: 2079-2092.

[62] Belimov AA, Kojemiakov AP, Chuvarliyeva CV. Interaction between barley and mixed cultures of nitrogen fixing and phosphate-solubilizing bacteria. *Plant Soil.* 1995; 173: 29-37.

[63] Wani PA, Khan MS, Zaidi A. Synergistic effects of the inoculation with nitrogen-fixing and phosphate-solubilizing rhizobacteria on the performance of field-grown chickpea. *J. Plant. Nutr. Soil Sci.* 2007; 170:283-287.

[64] Yi Y, Huang W, Ge Y. Exopolysaccharide: a novel important factor in the microbial dissolution of tricalcium phosphate. *World J. Microbio. and Biotech.* 2008; 4:1059-1065.

[65] Iniguez AL, Robleto EA, Kent AD, Triplett EW. 2004. Significant yield increase in *Phaseolus vulgaris* obtained by inoculation with a trifolitoxin-producing, Hup^+ strain of *Rhizobium leguminosarum* bv. phaseoli. *Online Crop Management.* 2004; doi:10.1094.

[66] Rodríguez H, Gonzalez T, Selman G. Expression of a mineral phosphate solubilizing gene from *Erwinia herbicola* in two rhizobacterial strains. *J. Biotech.* 2000; 84155-161.

In: Soil Fertility
Editors: D. P. Lucero and J. E. Boggs
ISBN: 978-1-60741-466-7

Chapter 2

MODEL SIMULATION OF CARBON AND NITROGEN DYNAMICS IN A BIOWASTE COMPOST FERTILISATION EXPERIMENT

***B. Foereid*[*a,d], *E. Erhart*[b], *E. Schmid*[c] and *W. Hartl*[b]**

[a] Plant & Soil Science, Cruickshank Building,
School of Biological Science, St Machar Drive, AB24 3UU.
[b] Ludwig Boltzmann-Institute for Biological Agriculture and Applied Ecology,
Rinnboeckstrasse 1,A-1110 Vienna, Austria
[c] University of Natural Resources and Applied Life Sciences Vienna;
Feistmantelstrasse 4A-1180 Vienna, Austria
[d] The Macaulay Institute, Craigiebuckler, Aberdeen AB15 8QH, U.K.

ABSTRACT

Municipal solid wastes are an increasing problem worldwide. The most environmentally friendly way of disposing of the organic fraction is to compost it and use it as soil-improvement in agriculture, as it both disposes of the waste and increases soil fertility. An experiment has been running for 11 years outside Vienna comparing bio-waste compost with mineral fertilisation and no fertilisation. The site is characterised by high natural fertility and dry climate in the growing season. Several soil and crop parameters were measured during the experiment, including yield, mineral nitrogen and organic carbon and nitrogen. Simulation models provide an excellent tool to integrate such data and provide an overall understanding of the system. We used this data set to parameterize and compare two simulation models, EPIC and CENTURY. These models share essentially the same carbon turnover model, but differ in the detail they describe crop growth and water balance. We found that EPIC predicted crop yield acceptably, while CENTURY underestimated it. Both models could predict total soil carbon and nitrogen reasonably well, despite large differences in predicted plant production. Both models predicted larger peaks of mineral nitrogen than observed and CENTURY overestimated and EPIC underestimated it overall. Neither of the models could predict soil water content very accurately. The models failed to account adequately for the slow-

* Tel: 01224 272257; Fax: 01224 272703; e-mail: b.foereid@abdn.ac.uk

release fertiliser effect of compost, possibly because of failure to accurately simulate effect of low water availability and litter quality on decomposition rate.

Keywords: Model comparison, Municipal waste, Soil carbon, Soil mineral nitrogen, Soil water content.

Introduction

The increasing amount of municipal solid wastes (household wastes) represents a growing problem in Europe and worldwide. The European Waste Management policy [European Commission, 1975] aims at prevention and, if this is not possible, recycling of wastes. Organics make up one third of municipal solid wastes. Composting of separately collected organic household wastes is one of the most environmentally friendly solutions for organic wastes. Household wastes can be composted together with garden wastes. Applying the produced biowaste compost on agricultural land provides nutrients for crop growth and improves soil tilth and fertility, and reduces the need for mineral fertilisers.

A field experiment comparing the effects of biowaste compost and mineral fertilisers on crop nitrogen uptake and quantity and quality of yield has been running for 11 years near Vienna, Austria. The results so far indicate that compost act as a slow release fertiliser in this relatively dry climate, with small effect initially, but increasing over time [Erhart *et al.,* 2005]. The amount of nitrogen leached from the treatments that received compost was modest [Erhart *et al*., 2007], which indicates that nitrogen may be stored in soil organic matter. That indicates that carbon was also stored as soil organic matter (SOM), and the system acted as a carbon sink. We assumed that the slow mineralization is related to recalcitrant compost material, mainly due to its content of woody garden waste and the dry climate on the site.

Because of concerns about global environmental change [King, 2004], the possibility of increasing the sequestering of carbon in terrestrial ecosystems has received considerable attention recently [Schlesinger, 1999; Smith *et al.,* 2000; Smith, 2004; Smith *et al.,* 2007]. A high level of SOM is also beneficial for soil fertility and soil structure [Christensen & Johnston, 1997; Sylvia *et al.,* 1998]. It is, therefore, of interest to adopt agricultural practices that can increase soil organic matter content in agricultural soils. Addition of organic matter to the soil increases soil carbon content over time, but the magnitude of this effect is variable [Christensen & Johnston, 1997]. The main problem in studies of soil carbon sequestration is the long time scale involved. Under temperate conditions it takes 100 years or more to reach a new equilibrium when management is changed [Christensen & Johnston, 1997]. One way to get a fairly accurate picture where insufficient data is available is to use well-validated dynamic models [Smith *et al.,* 2000; Foereid & Høgh-Jensen, 2004].

Using models is also an excellent way to integrate information from field experiments, particularly when many parameters have been measured as in this experiment. Soil carbon has not been extensively measured in the compost trial, and the time the experiment has been running is also relatively short. However, many other parameters related to element cycling were measured and they can be used to validate models of carbon and nitrogen dynamics. We compared two models for carbon and nitrogen turnover, CENTURY and EPIC. Both these models share essentially the same soil organic matter model [Izaurralde *et al.,* 2006], but differ in the level of detail they describe crop growth and water balance that give inputs to the carbon turnover model [Parton *et al.,* 1987; Williams, 1990]. Whilst CENTURY has been widely tested, the soil organic matter model is a recent addition to EPIC. It was therefore interesting to compare them on the same data set.

The purpose of the study was 1) to compare the two models using a good data set 2) to try to understand nutrient turnover in the system and predict carbon sequestration or loss in the future using process-based models and the available data.

MATERIAL AND METHODS

Experimental Design

The experiment is located close to Vienna (16° 30' E 48° 10' N), and includes three treatments with compost fertilisation, three treatments with mineral nitrogen fertilisation, five treatments with combined fertilisation and an untreated control (table 1). The five mixed treatments include the lowest level of mineral fertiliser with all three levels of compost, and the lowest level of compost with all levels of mineral fertiliser. The untreated control did not receive any mineral fertiliser or compost. There are six replications of each treatment arranged in a Latin rectangle design. Compost was applied in the autumn, whereas mineral fertiliser was applied in spring, except in 1998 when both were applied in spring. Table 1 shows the amounts of compost and mineral N fertiliser used at each application, as well as the relevant chemical characteristics measured at each application. The treatments will be referred to as in table 1 throughout, three levels of compost: C1 (lowest), C2 and C3 (highest), three levels of mineral fertiliser: N1 (lowest), N2 and N3 (highest) and combined treatments as N1C1, N1C2, N1C3, N2C1 and N3C1. The untreated control is referred to as 0 fert.

Table 1. Amounts of mineral nitrogen in kg ha^{-1} added at each application in the mineral fertiliser treatments (top). Measured compost chemical parameters at each application (middle) and amounts of compost applied per application in tones ha^{-1} of fresh matter (bottom). See text for explanation of the treatments

	Years						
	92-93	93-94	94-95	95-96	98	00-01	02
Mineral N treatment (kg ha^{-1})							
N1	25	40	60	30	60	30	30
N2	50	80	90	45	90	45	45
N3	75	120	120	60	120	60	60
Compost quality							
Water (%)	24	37	12	37	28	38.2	
C/N	21	19	23	49	9	19	
Lignin (%)	*	*	*	*	17	10.6	
Compost treatment (tones ha^{-1})							
C1	15	15	30	15	10	10	
C2	30	30	45	30	20	20	
C3	45	45	60	45	30	30	

* not measured.

Site and Field Management

The site has a Molli-gleyic Fluvisol and the climate was relatively dry (552 mm on average, potential evapotranspiration 1000 mm). The soil parameters used are given in table 2. Information on the soil profile was taken from Erhart *et al.* [2002]. The crop rotation used in all treatments consisted of mostly winter cereals with potatoes about every fourth year (table 3). Nitrogen deposition at the site was 20 kg ha^{-1} [Puxbaum & Gregori 1998]. All treatments had the same crop rotation and the same field management.

CENTURY Model

The CENTURY model has been well described elsewhere [Parton *et al.,* 1987; 1988; 1994]. CENTURY was originally developed for long term simulation of grasslands on the Great Plains of the US [Parton *et al.,* 1987]. The model was one of the best-performing in a comparison of soil organic matter models [Smith *et al.,* 1997]. In brief, the model simulates the dynamics by using three pools of soil organic matter and two pools of organic litter. The model also contains a crop growth sub-model. The model operates with time-steps of one month with precipitation and temperature as climate inputs. We used version 5 of the model, and nitrogen was considered the only limiting nutrient.

EPIC Model

The EPIC model (erosion productivity impact calculator) was developed by an USDA modeling team in the early 1980's to assess the status of U.S. soil and water resources [Williams *et al.,* 1984; Williams, 1990; Jones *et al.,* 1991]. The first major application of EPIC was to evaluate soil erosion impact for 135 U.S. resource regions [Putman *et al.,* 1988]. The model has later been expanded to allow simulation of many processes important for agricultural land management [Sharpley & Williams, 1990; Williams, 1995; 2000]. Recently a model of carbon and nitrogen turnover based on the equations in CENTURY has been introduced [Izaurralde *et al.,* 2001; 2006]. The routines were modified somewhat to fit into the structure of EPIC. See Izaurralde et al. [2006] for details. EPIC has a more detailed crop model, time steps of one day and a more detailed water model than CENTURY. It also requires a detailed soil profile and can simulate up to 15 layers. Of the six evapotranspiration equations available, we used the Penman-Monteith equation. In the CENTURY model, the effect of cultivation was made to last for two months, when cultivation was done in spring or autumn (between September and April) as the effect has been shown to last longer in cool climate (Keough, pers comm.).

Weather Data

Precipitation was measured at the site from 1996-2001. Before that, data from the meteorological station at Grossenzersdorf (Zentralanstalt für Meteorologie und Geodynamik, Wien) 4 km northeast of the experimental site were used. The temperature data are from the

Table 2. Soil parameters for the soil profile (Erhart *et al.* 2002)

Horizon depth	Clay	Silt	Sand	Coarse particles (> 2 mm)	Humus	$CaCO_3$	Bulk density	pH	Saturated conductivity	Water content at field capacity (300 hPa)	Water content at Permanent wilting point (15000 hPa)
cm	(%)	(%)	(%)	(%)	(%)	(%)	g cm^{-1}		m d^{-1}	% vol	%vol
0-10	22	61	17	0	3.79	27.5		7.4			
10-25	24	59	17	<1/7	4.99	26.5	1.16	7.4	1.82	34.6	14.1
25-35	22	61	17	<1/6	3.59	27.2	1.16	7.4	4.08	31.9	13.1
35-50	20	61	19	0	1.50	29.9		7.6			
50-60	2	68	30	<1/15	1.20	29.0	1.43	7.7	1.71	27.5	8.9
60-70	15	47	38	0	0.80	26.6		7.7			
70-80	17	61	22	0	0.90	31.0	1.40	7.7	0.51	34.2	13.4
80-90	19	70	11	0	1.00	36.5		7.8			
90-100	4	23	73	0	0.40	27.1	1.32	7.7	2.4	12.9	3.5
100-110	4	34	62	0	0.50	29.0		7.7			
110-125	9	64	27	0	0.50	30.0	1.43	7.8	0.23	35.6	6.4
125-140	1	15	84	0	0.30	24.5	1.37	7.7	3.83	7.1	1.0
140-150	2	18	80	0	0.30	24.7		7.7			
150-160	2	20	78	0	0.30	26.1		7.6			
160-170	1	16	83	0	0.20	24.4		7.6			
170-180	0	7	93	0	0.20	21.8	1.44	7.6	6.13	4.9	1.1
180-190	2	13	85	0	0.30	23.0		7.7			
190-205	7	44	49	0	0.40	27.1	1.43	7.7	0.48	20.1	3.9
205-210	4	16	80	0	0.30	15.2		7.6			
210-220	1	2	97	0	0.20	12.7	1.44	7.6	14.86	5.9	0.5
220-230	1	5	87	7/33	0.20	13.6		7.5			

Table 3. Crop grown and field management each year. Crop grown, planting and harvest month, month of compost and fertiliser application and cultivation are shown. Note that when winter crop was planted the year before, that month is listed in the year of main crop growth

Year	Crop	Planting Month	Harvest month	Fertiliser/compost month	Cultivation/ month
1992	Winter rye	09-91	07-92	Compost: 09	Cultivation / 07 Ploughing / 09 Hoeing / 04
1993	Winter rye	09-92	07-93	Compost : 09 Fertiliser: 03	Cultivation / 07 Ploughing / 09 Hoeing / 04
1994	Potatoes	04-94	08-94	Compost: 09 Fertiliser: 04	Ploughing / 09 Hoe potato / 03 + 04 Manual weed control / 5
1995	Winter wheat	10-94	07-95	Compost: 09 Fertiliser: 03+05	Cultivation / 07 Ploughing / 09
1996	Oats	04-96	08-96		Cultivation / 09 Ploughing / 09 Manual weed control / 05 Hoeing / 05
1997	Spelt	10-96	08-97	-	Cultivation / 8 Ploughing / 10 Manual weed control / 04 Hoeing / 04 + 05
1998	Early potatoes	03-98	07-98	Compost: 04 Fertiliser: 03	Cultivation / 03 + 07 + 08 +09 Ploughing / 09 Hoe potato / 03 + 04 Manual weed control / 05
1999	Winter wheat	10-98	07-99	-	Ploughing / 09 Hoeing / 3
2000	Winter barley	09-99	06-00	Compost: 09	Cultivation / 7 + 8 Ploughing / 09 Hoeing / 04
2001	Winter rye	09-00	07-01	Fertiliser: 03	Cultivation / 07 + 08 Ploughing / 09 Hoeing / 04
2002	Winter wheat	10-01	07-02	Fertiliser: 04	Cultivation / 07 Ploughing / 09 Hoeing / 04 + 04

same meteorological station for the whole period. The EPIC model requires solar radiation as input, and as this was not available, it was estimated based on temperature and latitude.

The formula of Bristow & Campbell [1984] modified by Donatelli & Campbell [1998] was used. A simple sensitivity test showed that changing radiation by 10 % changed crop production by 5-6 %, showing that small errors in the radiation calculation will not have large effects on simulation outcome. Both models had stochastic weather generators, which were used during initialisation.

Soil Water

Soil water content was measured at the site by TDR (Time Domain Reflectometry, Soilmoisture Equipment Corp.; U.S.A.) at five depths at three points, one in the control treatment, one in the N3 treatment and one in the C3 treatment [Erhart *et al.,* 2003]. At the site, soil water could rise by capillary forces and supply the crop with some water during dry periods [Erhart *et al.,* 2003]. As water can only move downwards in the CENTURY model [Paustian *et al.,* 1992; CENTURY 5 User's Guide and Reference], we supplied water in the model by irrigation so that water content at 150 cm soil depth matched observed data. The EPIC model can simulate upwards movement of water. Maximum and minimum groundwater table measured at the site was used as inputs. Simple sensitivity test of the effect of changing water holding capacity and irrigation on yield and mineral nitrogen content were performed

Biowaste Compost Parameters

Dry matter, carbon and nitrogen content were measured in the compost at every application, and these values were used in the simulation. The chemical and physical properties of compost were analyzed according to OENORM S 2023 [1993]. N_{tot} was calculated as a sum of N determined by the Kjeldahl method plus NO_3^--N determined photometrically. For C_{org}, the loss on ignition was determined in a muffle furnace and the value multiplied by 0.58 to obtain C_{org}. Lignin content was measured with the method of van Soest [1963], but only at the last two applications (table 1). Values from the first measurement were therefore used in the CENTURY simulation of all the applications before that. EPIC does not take lignin content as input in organic fertiliser amendments.

Crop Production

Crop yield as marketable product was measured every year. To compare with CENTURY model output given as g C m^{-2}, it was assumed that 45% of dry matter was carbon. Until 1997, nitrogen contents of crops and straw were estimated using standard values [Früchtenicht *et al.,* 1993]. Later, total N in grain and potato tubers was measured using a modified Kjeldahl procedure [Horneck & Miller, 1998].

Organic Carbon and Nitrogen Content in the Soil

Total carbon and nitrogen content was measured just after the start of the experiment, in spring 1993, and during the experiment in 1997 and 2003. In 1993 only the control plots (no addition) were sampled, but all treatments were sampled on the other sampling dates. At each sampling date eight or more cores (2 cm in diameter) down to 30 cm were composited from each plot. Total nitrogen was measured following the Dumas method with an automated instrument [Bremner & Mulvaney, 1982]. Total carbon was determined by dry combustion [Nelson & Sommers, 1982]. For organic carbon, the amount of carbonate, analyzed by the modified Van Slyke Manometric method [Nelson, 1982] was subtracted from the total carbon.

Mineral Nitrogen

Mineral nitrogen was measured three times a year (February-March, June-July and September -October) from 1993 to 2002. At each sampling date eight cores (2 cm in diameter) down to 30 cm were bulked from each plot. Nitrate was analyzed according to Merck [1986]. Ammonium-N in the soil was measured on selected sampling dates and was usually < 5 kg ha^{-1}. We therefore considered ammonium-N to be negligible at the experimental site. These data were also compared with the model's mineral nitrogen output values.

Parameter Adaptation

Both models have been applied with default parameter settings to allow fair comparisons between model performances. The soil parameters used for model calibration and those used for model evaluation were measured independently of each other. We therefore get a measure of how well the models may perform when they are eventually used in regional analyses [e.g. Burke *et al.,* 1991; Tan & Shibasaki, 2003]. The EPIC model requires heat units (degree days) as inputs. That was found in Broons-Prins et *al.* [1993], and then calibrated so that the crop was mature at harvest time.

RESULTS

Crop Yield and Nitrogen Content

The observed yields were much larger than those predicted by CENTURY in most years (figure 1, table 4). The predictions of EPIC were closer, but still only explaining a bit over 50% of the total variance (table 4), and the predictions also were on the low side (figure 1).Also nitrogen content in the yield was severely underpredicted by CENTURY in most years (figure 1). There was little bias in EPIC's predictions (figure 1; table 4), so that over 70% of the total variance was explained.

Table 4. Coefficients of determination (r^2) for the two models and average and standard deviation for the data and models

data	DATA	CENTURY	EPIC
R^2			
Crop drymatter yield		0.287	0.536
Nitrogen in yield		0.391	0.736
Soil carbon		0.975	0.922
Soil nitrogen		0.962	0.780
Mineral nitrogen		0.499	0.499
Soil water content – 10-35 cm		0.500	0.829
Soil water content – 60 cm		0.887	0.388
Average			
Crop drymatter yield	4319	1597	2339
Nitrogen in yield	76.1	30.4	60.7
Soil carbon	2.07	2.42	1.63
Soil nitrogen	0.226	0.278	0.148
Mineral nitrogen	46.58	127.49	12.24
Soil water content – 10-35 cm	32.49	22.86	24.12
Soil water content – 60 cm	26.4	30.9	13.25
Standard deviation			
Crop drymatter yield	614.6	1263.6	1456
Nitrogen in yield	23.3	34.4	38.5
Soil carbon	0.135	0.262	0.0967
Soil nitrogen	0.0127	0.0217	0.00508
Mineral nitrogen	32.7	132.8	30.9
Soil water content – 10-35 cm	5.36	6.74	6.23
Soil water content – 60 cm	4.63	10.63	6.45

Soil Water

Neither model simulated water content accurately (table 4). Both models did better in the top layer than in the lower soil layer, and there were also small differences between the three plots measured (figure 2). EPIC underestimated water content particularly at 60 cm depth, while CENTURY predicted large variations (figure 2; table 4). As capillary may explain this to some extent, the ground water table depth is shown in figure 2 as well. A simple sensitivity test of EPIC showed that increasing clay content in the soil increased water content, but had little effect on yield and soil mineral nitrogen content. A sensitivity test of CENTURY showed that increasing water availability through irrigation did not increase yield substantially

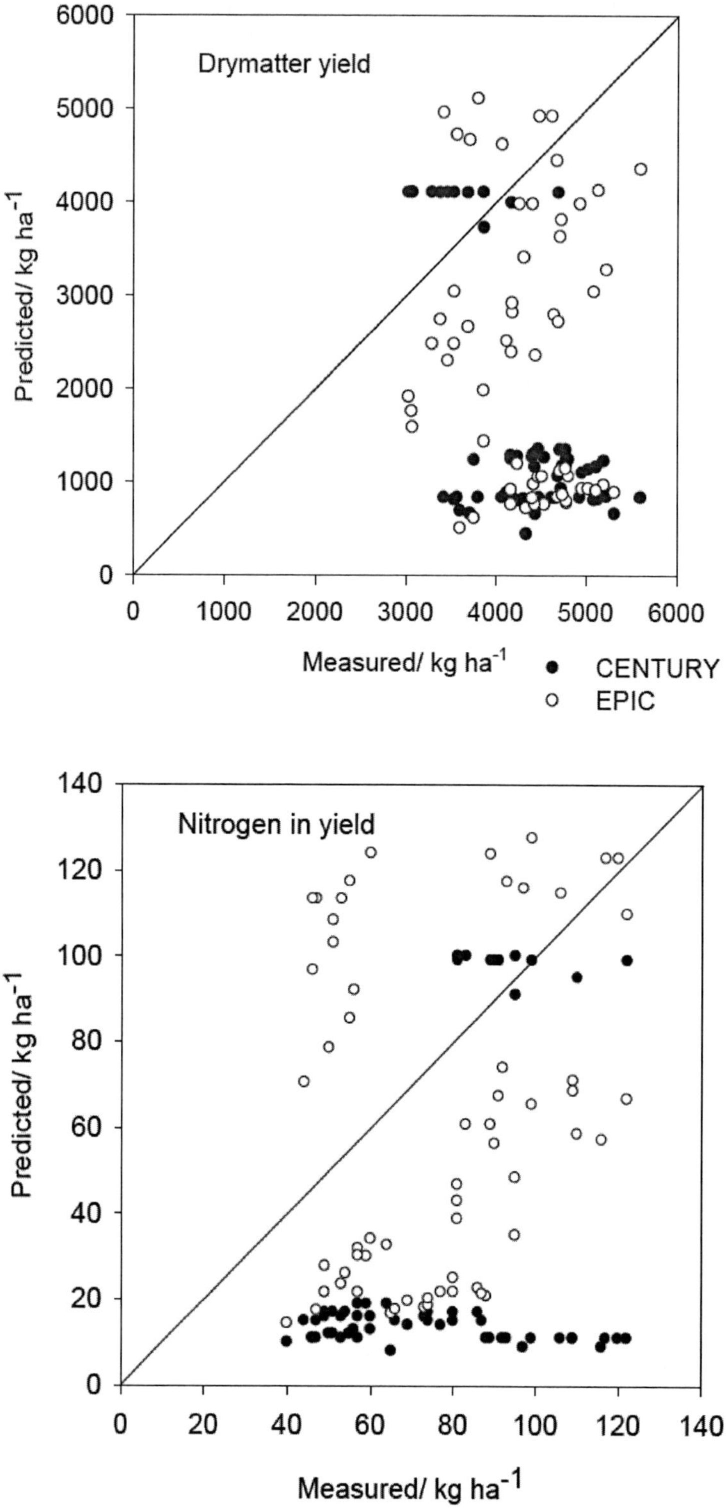

Figure 1. Measured and simulated crop yield and total nitrogen in grain for CENTURY and EPIC. Diagonal shows line of perfect fit.

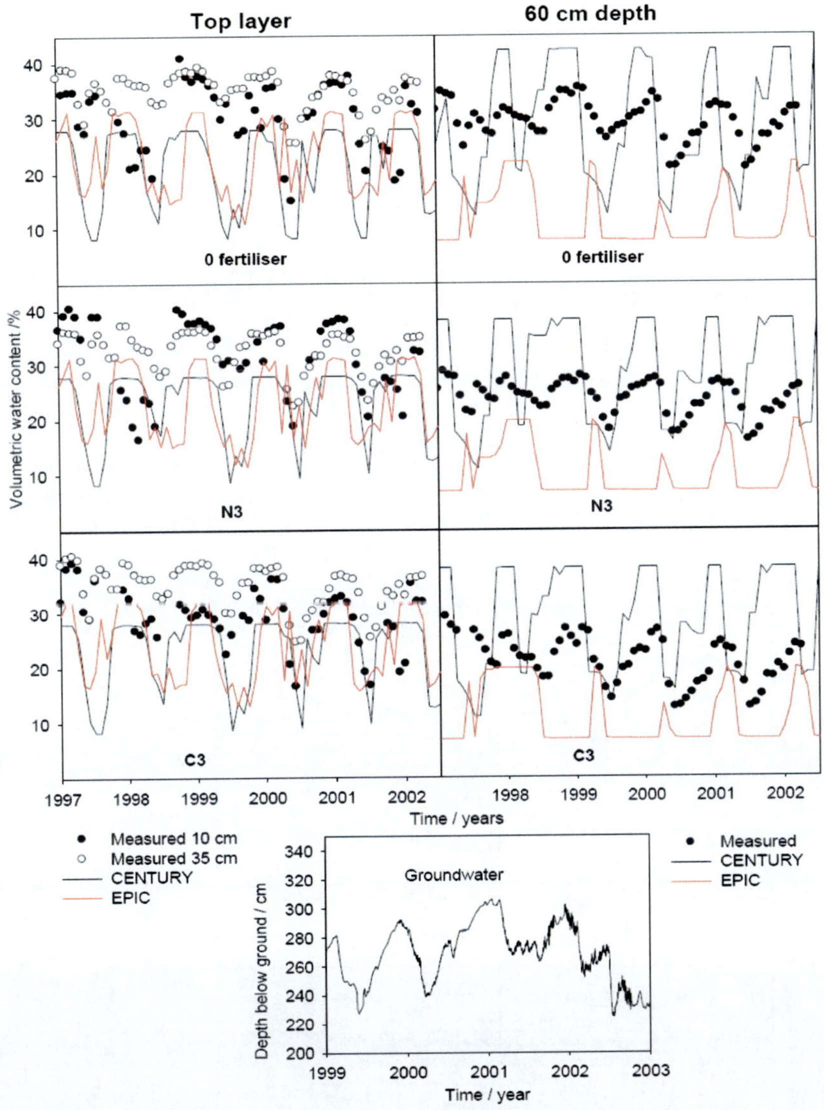

Figure 2. Soil water content (% vol.), measured and simulated values. For comparison, measured groundwater table in cm below the surface is shown below. The measurements at 10 and 35 cm were compared to simulated values for the 0-30 cm layer (left), while the measured values at 60 cm depth were compared to the simulated values for the 50-70 cm layer (right).

Organic Carbon and Nitrogen

Both models simulated total soil carbon and to a lesser extent nitrogen adequately when measured against the limited measured data available (table 4). However, while CENTURY predicted little change in soil carbon without compost and an increase in the treatments receiving compost, EPIC predicted decline in all treatments, and was closer to the measured data (figure 3). CENTURY overestimated soil mineral nitrogen content, and both models predicted large peaks (in the case of CENTURY sometimes even negative ones), and EPIC underestimated soil mineral nitrogen somewhat (table 4, figure 4).

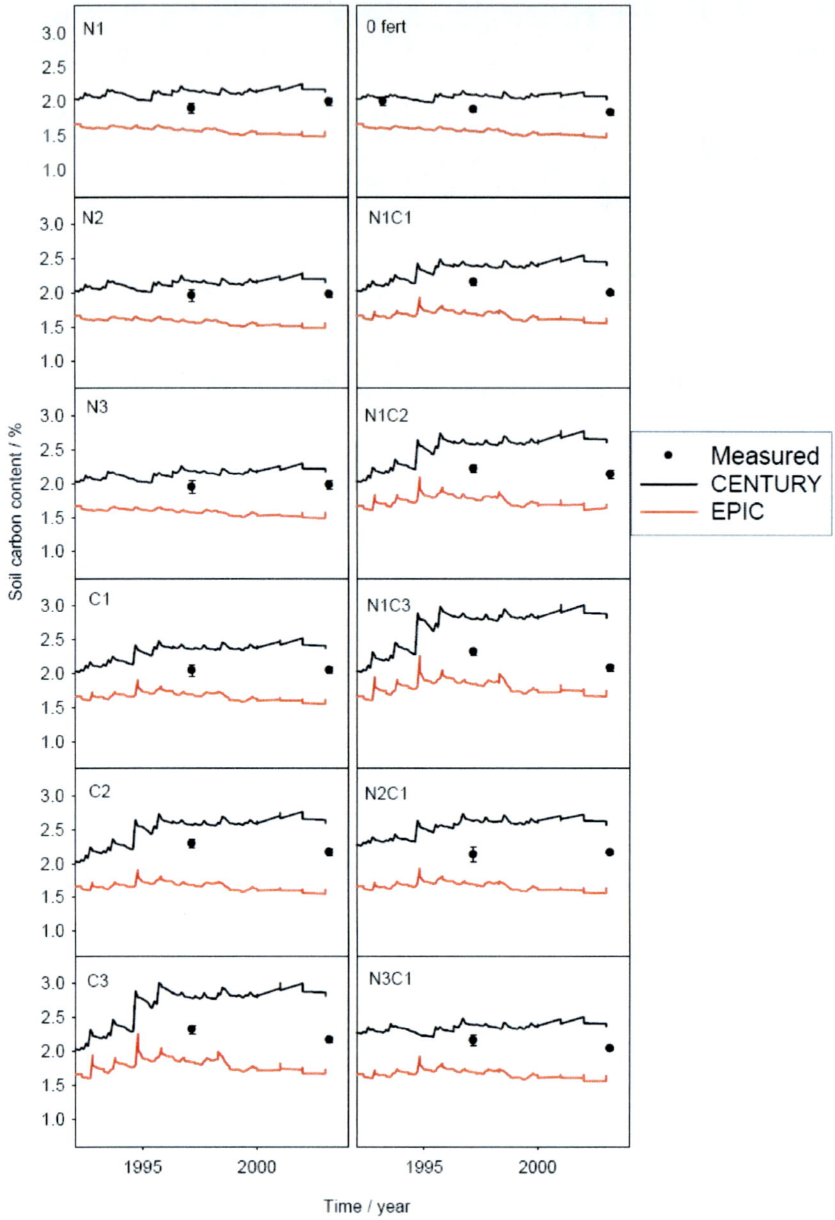

Figure 3. Soil total organic carbon (%) in the top 30 cm measured and simulated with CENTURY and EPIC as a function of time for all the treatments. Error bars are standard error.

Mineral Nitrogen

The data indicate that mineral nitrogen remained at the same level in the latter part of the experiment, when fertiliser and compost additions became more infrequent. Neither model accounted adequately for this, particularly not EPIC, which predicts almost no mineral nitrogen in the latter part of the experiment (figure 4). Overall EPIC had both larger respiration losses and smaller residue additions than CENTURY.

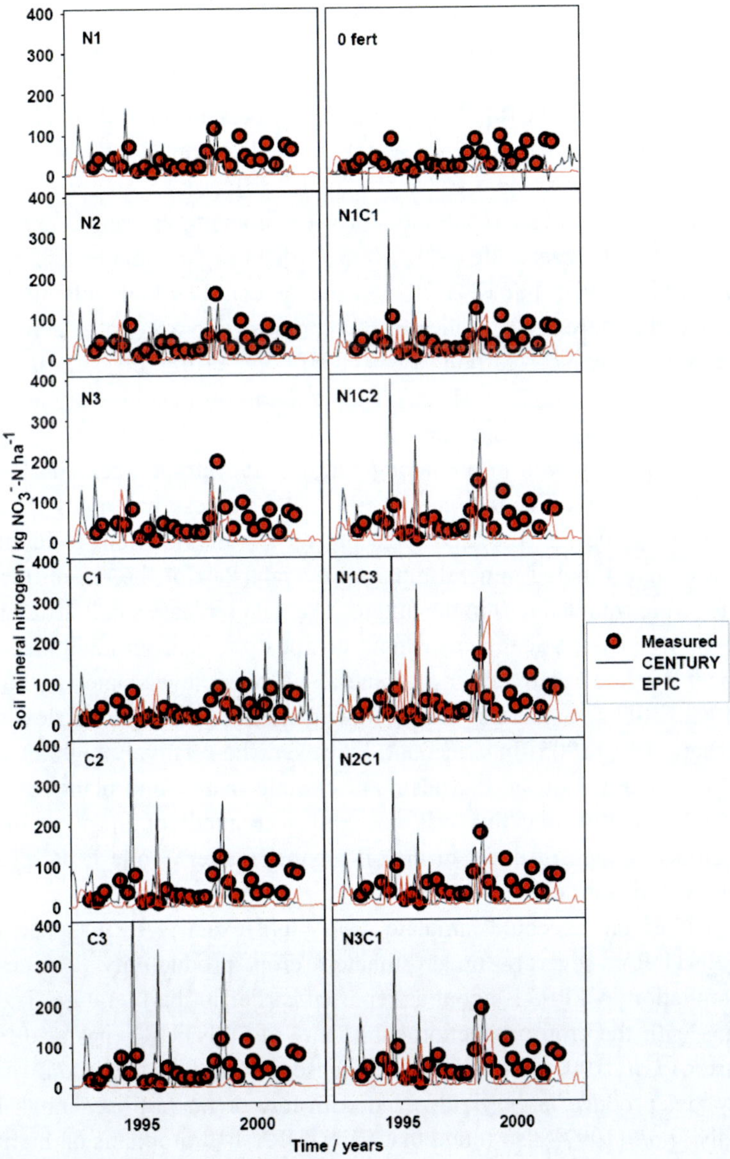

Figure 4. Soil mineral nitrogen levels (kg N ha^{-1}) in the top 30 cm measured and simulated with EPIC and CENTURY as a function of time for all the treatments.

DISCUSSION

Both models predicted soil organic carbon and nitrogen reasonably, but CENTURY tended to overestimate increase in soil carbon, in particular it overestimated the effect of compost. Also other authors have found that CENTURY overestimates the effect of organic inputs [Paustian *et al.,* 1992; Parton & Rasmussen, 1994; Foereid & Høgh-Jensen, 2004], which suggests that CENTURY may also suffer from underestimation of plant input balanced by too low decomposition rates [Bruun *et al.,* 2003]. This problem may be more acute at high fertility sites, like this one. This is supported by the fact that CENTURY could simulate soil organic carbon reasonably and even overestimate it, although plant production, and plant carbon input to the soil, was severely underestimated. Although EPIC simulated crop yield much more closely than CENTURY, EPIC predicted less plant input than CENTURY. EPIC therefore simulated decrease in soil organic matter in all treatments, despite being based on the same soil organic matter as CENTURY. The differences are probably caused by differences in the predicted plant input and soil water content, as well as differences in the layer structure of the soil [Izzauralde et al., 2006]. Added organic matter (compost) also had much less effect in EPIC than in CENTURY. A reason could be that EPIC does not take the lignin content, and therefore recalcitrance of the compost into account. There is not enough data to ascertain whether soil carbon was actually increasing or decreasing during the experimental period, but a rather small effect on soil carbon of added organic material and fertiliser is certainly supported by the data.

Neither model did very well in predicting soil mineral nitrogen content. However, large and sometimes even negative fluctuations in CENTURY suggest that time steps of one month are inadequate to describe fast processes like nitrogen mineralization. EPIC also predicted large peaks, but underestimated mineral nitrogen overall. Part of the reason for that may be that small values were rounded to 0 in the output. The data indicate small differences between treatments and between years with and without compost or fertiliser additions. Both models failed to account for the relatively high mineral nitrogen levels late in the experiment, particularly EPIC. EPIC does not take the lignin content and therefore the slow release effect of compost [Erhart *et al.,* 2005] into account. However, the relatively high levels of mineral nitrogen in all treatments indicate that also soil organic matter and plant residues acted as source of mineral nitrogen, and this was not well accounted for in either of the models. CENTURY predicts storage of nitrogen in soil organic matter, while EPIC predicts loss of nitrogen through volatilisation.

Although neither model could simulate crop yield exactly, EPIC's performance was acceptable. CENTURY severely underestimated crop productivity. Accuracy of crop productivity simulation in EPIC is about as reported earlier in the literature [Izzuralde *et al.,* 2006]. Problems with the crop production sub-model of CENTURY has also been reported previously [Parton *et al.,* 1996; Foereid & Høgh-Jensen 2004]. The site is known to have high natural fertility and produce large yields even with low or no fertiliser input. Parton *et al.* [1993] and Kelly *et al.* [1997] also found that CENTURY had problems on high productivity sites. EPIC probably does better because it has more possibilities for site specific calibration (degree-days) and shorter time steps.

CENTURY model has previously been tested at a site in Hungary [Falloon & Smith, 2002; Falloon *et al.,* 2002] and fit to the data was fair. However, the model was tested against soil carbon data only; no plant growth or yield data were used. Kelly *et al.* [1997] tested CENTURY at two sites in Central Europe, Bad Lauchstadt and Prague. They found that CENTURY had limited success in predicting year to year variation in crop productivity, but that the average was close enough to predict soil carbon and nitrogen. There are few tests so far of the organic matter model of EPIC, and results suggest fair performance and that the model underestimates at high soil carbon and overestimates at low soil carbon [Izzauralde *et al.,* 2006].

Neither model predicted soil water content very accurately. Simulation was somewhat more accurate in some plots than in others. Part of the reason may be that the soil profile was described only one place in the field, whilst the thickness and to some extent the content of the various horizons varied slightly over the field. Particularly EPIC has a detailed water model, and correct profile information is important. Better predictions of soil water could be obtained by calibrating field capacity and wilting point from the observed soil water. As the purpose of this study was to compare the models using independently measured values, and see how good predictions we can expect, we have chosen not to do that. As the measured data at 60 cm never appear to stay at any minimum at an extended period of time, we think that it may not actually reach wilting point due to capillary rise. Although EPIC takes account of that to some extent, it appears to be far more important than the model can account for. However, incorrect predictions of soil water content were not enough to account for other discrepancies between measured and simulated results (yield and mineral nitrogen content) for either of the models. Both models predicted that most of the water was lost by evapotranspiration, which is probably realistic given the semi-arid climate at the site.

Conclusion

We can conclude that the effect of compost on soil organic matter was small, and this was accounted for better in EPIC than in CENTURY. Neither model could reproduce the slow release of mineral nitrogen from the added compost. Differences between the models predictions are due to differences in the calculations of plant productivity, soil water content and soil profile description. Neither model could satisfactorily simulate all the measured variables, even when a lot of detailed site specific input data were available. The site is dry with but with water rising with capillary rice and high natural fertility. These may be factors that the models are poorly equipped to deal with. The results indicate that we do not have a mechanistic understanding of the carbon and nitrogen dynamics at the site. To extrapolate far into the future may therefore not be justified.

Acknowledgment

The compost experiment was supported by funds from the City of Vienna. The simulation work was supported by the Scottish Executive Environment and Rural Affairs Department and a travel grant from EFS program on *The Role of Soils in the Terrestrial Carbon Balance*. Advice from Cindy Keough at Colorado State University on the use of the CENTURY model is gratefully acknowledged.

REFERENCES

Boons-Prins E.R., Koning G.H.J., van Diepen C.A., & Penning de Vries F.W.T. (1993) *Crop specific simulation parameters for yield forecasting across the European Community.* Simulation Reports CABO-*TT*, no. 32, DLO, Wageningen Agricultural University

Bremner J., & Mulvaney C. Nitrogen - Total. In: Page, A.L. Miller, R.H. Keeney, D.R.

Editors. *Methods of Soil Analysis: Part 2.*, American Society of Agronomy, Madison, Wisconsin. 1982. pp. 595-624.

Bristow, K.L., & Campbell, G.S. (1984). On the relationship between incoming solar radiation and daily maximum and minimum temperature. *Agricultural and Forest Meteorology, 31,* 159-166.

Bruun, S., Christensen, B.T., Hansen, E.M., Magid, J. & Jensen, L.S. (2003). Calibration and validation of the soil organic matter dynamics of the Daisy model with data from the Askov long-term experiments. *Soil Biology & Biochemistry, 35,* 67-76.

Burke I.C., Kittel T.G.F., Lauenroth W.K. Snook C.M. & Parton W.J. (1991). Regional Analysis of the Central Great Plains. *BioScience, 41,* 685-692.

Christensen, B.T., & Johnston, A.E. Soil organic matter and soil quality - lessons learned from long- term experiments at Askov and Rothamsted. In: Gregorich E.G. Carter M.R. editors. *Soil Quality for Crop Production and Ecosystem Health,* 1997. Amsterdam, Elsevier. pp. 399-430.

Donatelli, M., & Campbell, G.S. (1998). A simple model to estimate global solar radiation. 5th Congeress of the European Society for Agronomy, Nitra, Slovakia.

Erhart E., Feichtinger F., Hartl W., Huspeka C., & Scheidl, A. (2002). Die Auswirkung der Kompostanwendung im Biologischen Landbau auf die Nitratauswaschung im Vergleich zu mineralischer Düngung. Endbericht. Forschungsvorhaben im Auftrag des Bundesministeriums für Land- und Forstwirtschaft, Umwelt und Wasserwirtschaft und der Magistratsabteilung 48 der Stadt Wien.

Erhart E., Hartl W. & Feichtinger F. (2003). Nährstoffflüsse im Ackerbau bei Kompostdüngung im Vergleich mit mineralischer Düngung - Ergebnisse der Lysimeteranlage Lobau. in: *Bundesanstalt für alpenländische Landwirtschaft Gumpenstein des BMLFUW (Hrsg.)*: Bericht über die 10. Gumpensteiner Lysimetertagung, Gumpenstein, 29.-30. 4. 2003. pp. 71-75.

Erhart, E., Hartl, W. & Putz, B. (2005). Biowaste compost affects yield, nitrogen supply during the vegetation period and crop quality of agricultural crops. *European Journal of Agronomy, 23,* 305-314.

Erhart E., Feichtinger F. & Hartl W. (2007). Nitrogen leaching losses under crops fertilized with biowaste compost compared with mineral fertilization. *Journal of Plant Nutrition and Soil Science, 170,* 1-7.

European Commission 1975. Council Directive 75/442/EEC http://europa.eu.int/smartapi/cgi.

Falloon, P. & Smith, P. (2002). Simulating SOC changes in long-term experiments with RothC and CENTURY: model evaluation for a regional scale application. *Soil Use and Management, 18,* 101-111.

Falloon, P., Smith, P., Szabo, J. & Paztor, L. (2002). Comparison of approaches for estimating carbon sequestration at the regional scale. *Soil Use and Management, 18,* 164-174.

Foereid, B. & Høgh-Jensen, H. (2004). Carbon sequestration potential of organic agriculture in northern Europe - a modelling approach. *Nutrient Cycling in Agroecosystems, 68,* 13-24.

Früchtenicht K., Heyn J., Kuhlmann H., Laurenz L. & Müller S. (1993). Pflanzenernährung und Düngung. In: Hydro-Agri Dülmen Editor. *Faustzahlen für Landwirtschaft und Gartenbau* Münster-Hiltrup, Aufl. Landwirtschaftsverlag. pp. 254-295.

Horneck D., & Miller R. Determination of total nitrogen in plant tissue. in: Kalra Y P Editor.

Handbook of reference methods for plant analysis Boca Raton, Florida. CRC Press, 1998; pp. 75-83.

Izaurralde, R.C., Williams, J.R., McGill, W.B. & Rosenberg, N.J. (2001). Simulating soil carbon dynamics, erosion, and tillage with EPIC. Washington D.C., First National Conference on Carbon Sequestration.

Izaurralde, R.C., Williams, J.R., McGill, W.B., Rosenberg, N.J. & Quiroga Jakas, M.C. (2006). Simulating soil C dynamics with EPIC: Model description and testing against long-term data. *Ecological Modeling*, *192*, 362-384.

Jones, C.A., Dyke, P.T., Williams, J.R., Kiniry, J.R., Benson, V.W. & Griggs, R.H. (1991). EPIC: an operational model for evaluation of agricultural sustainability. *Agricultural Systems*, *37*, 341-350.

Kelly, R.H., Parton, W.J., Crocker, G.J., Grace, P.R., Klor, J., Korchens, M, Poulton, P.R. & Richter, D.D. (1997). Simulating trends in soil organic carbon in long-term experiments using the CENTURY model. *Geoderma,* 81, 75-90.

King, D.A. (2004). Climate change science: adapt, mitigate, or ignore? *Science*, *303*, 176-177.

Merck Hrsg. (1986). *Nitrat Schnelltest Handbuch*. Darmstadt, Firmenschrift E. Merck.

Nelson R.E. Carbonate and Gypsum. In: Page A.L. Miller R.H. Keeney D.R Editors. *Methods of soil analysis. Part 2.* Madison, WI. American Society of Agronomy, Soil Science Society of America,1982. pp. 181-197.

Nelson D., & Sommers L. Total carbon, organic carbon, and organic matter. In: Page A.L. Miller R.H. Keeney D.R Editors. *Methods of Soil Analysis. Part 2.* Madison, WI. American Society of Agronomy, Soil Science Society of America, 1982. pp. 539-579

OENORM S 2023. (1993). *Untersuchungsmethoden und Güteüberwaschung von Komposten.* Wien, Österreichisches Normungsinstitut.

Parton, W.J. & Rasmussen, P.E. (1994). Long-term effects of crop management in wheat-fallow: II. CENTURY Model Simulations. *Soil Science Society of America Journal, 58*, 530-536.

Parton, W.J., Schimel, D.S., Cole, C.V. & Ojima, D.S. (1987). Analysis of factors controlling soil organic matter levels in Great Plain grasslands. *Soil Science Society of America Journal, 51*, 1173-1179.

Parton, W.J., Stewart, J.W.B., & Cole, C.V. (1988). Dynamics of C, N, P and S in grassland soils: a model. *Biogeochemistry, 5*, 109-131.

Parton, W.J., McKeown, B., Kirchner, V. & Ojima, D.S. (1992). *CENTURY Users Manual.* Colorado State University, Fort Collins, Colorado. NREL Publication.

Parton, W.J., Scurlock, J.M.O., Ojima, D.S., Gilmanov, T.G., Scholes, R.J., Schimel, D.S., Kirchner, T., Menaut, J.-C., Seastedt, T., Moya, E.G., Kammalrut, A. & Kinyamrio, J.I. (1993). Observations and modeling of biomass and soil organic matter dynamics for the grassland biome worldwide. *Global Biogeochemical Cycles, 7*, 785-809.

Parton, W.J., Ojima, D.S., & Cole, C.V. A general model for soil organic matter dynamics: sensitivity to litter chemistry, texture and management. In: Bryant, R.B. & Arnold, R.W. Editors. *Quantitative Modeling of Soil Forming Processes* Soil Science Society of America Special Publication, Madison, WI. 1994. pp. 147-167.

Parton, W.J., Ojima, D.S. & Schimel, D.S. 1996. Models to Evaluate Soil Organic Matter Storage and Dynamics. In: Carter, M.R. & Stewart, B.A.Editors *Structure and Organic Matter Storage in Agricultural Soils* Boca Raton, Florida.CRC Press, pp. 421-448.

Paustian, K., Parton, W.J. & Persson, J. (1992). Modeling Soil Organic Matter in Organic-Amended and Nitrogen-Fertilized Long-Term Plots. *Soil Science Society of America Journal, 56*, 476-488.

Putman, J., Williams, J.R. & Sawyer, D. (1988). Using the erosion-productivity impact calculator (EPIC) to estimate the impact of soil erosion for the 1985 RCA appraisal. *Journal Soil Water Conservation, 43*, 321-326.

Puxbaum, H. & Gregori, M. (1998). Seasonal and annual deposition rates of suphur, nitrogen and chloride species to an oak forest in north-eastern Austria (Wolkersdorf, 240 m A. S. L.). *Atmospheric Environment, 32*, 3557-3568.

Schlesinger, W.H. (1999). Carbon Sequestration in Soils. *Science, 284*, 2095.

Sharpley, A.N., & Williams, J.R. Erosion/Productivity Impact Calculator: 1. Model Documentation. 1990. U.S. Dept. Agric. Tech. Bull. No. 1786.

Smith, P., Smith, J.U., Powlson, D.S., McGill, W.B., Arah, J.R.M., Chertov, O.G., Coleman, K., Franko, U., Frolking, S., Jenkinson, D.S., Jensen, L.S., Kelly, R.H., Klein-Gunnewiek, H., Komarov, A.S., Li, C., Molina, J.A.E., Mueller, T., Parton, W.J., Thornley, J.H.M. & Whitmore, A.P. (1997). A comparison of the performance of nine soil organic matter models using datasets from seven long-term experiments. *Geoderma, 81*, 153-225.

Smith, P., Powlson, D.S., Smith, J.U., Fallon, P. & Coleman, K. (2000). Meeting Europe's climate change commitments: quantitative estimates of the potential for carbon mitigation by agriculture. *Global Change Biology, 6*, 525-539.

Smith,P. (2004). Soils as carbon sinks: the global context. *Soil Use and Management 20*, 212-218.

Smith P., Martino D., Cai Z., Gwary D., Janzen H., Kumar P., McCarl B., Ogle S., O'Mara F., Rice C., Scholes B., Sirotenko O., Howden M., McAllister T., Pan G., Romanenkov V., Schneider U. & Towprayoon S. (2007). Policy and technological constraints to implementation of greenhouse gas mitigation options in agriculture. *Agriculture, Ecosystems and Environment 118*, 6-28.

Sylvia, D.M., Fuhrmann, J.J., Hartel, P.G., & Zuberer, D.A. (1998). Principles *and Applications of Soil Microbiology*. New Jersey, Prentice Hall.

Tan G. & Shibasaki R. (2003). Global estimation of crop productivity and the impacts of global warming by GIS and EPIC integration. *Ecological modelling, 168*, 357-370.

van Soest, P.J. (1963). Use of detergents in the analysis of fibrous feeds. *Journal of AOAC, 46*, 825-835.

Williams, J.R. 1995. The EPIC model. In: Singh V.P. Editor Computer Models of Watershed Hydrology, Colorado, *Highlands Ranch,* pp. 909-1000.

Williams, J.R. (1990). The erosion productivity impact calculator (EPIC) model: a case history. *Philosophical Transactions of the Royal Society of London, 329*, 421-428.

Williams, J.R., Arnold, J.G., & Srinivasan, R. The APEX model. BRC Report No. 00-06. 2000. Texas Agricultural Extension Service, Texas A&M University, Temple, TX: *Agricultural Experimental Station.*

Williams, J.R., Jones, C.A., & Dyke, P.T. (1984). A modeling approach to determining the relationship between erosion and productivity. In: *Transactions ASAE* pp. 129-144.

In: Soil Fertility
Editors: D. P. Lucero and J. E. Boggs
ISBN: 978-1-60741-466-35

Chapter 3

IMPACT OF SOIL FERTILITY ON RICE PRODUCTIVITY AND EFFICIENCY: A CASE STUDY FROM BANGLADESH

Sanzidur Rahman *

School of Geography, University of Plymouth, Plymouth, PL4 8AA, UK

ABSTRACT

Soil fertility is an important factor in enhancing crop productivity but is not commonly included in studies analyzing productivity and efficiency. In this study, we examine the influence of selected soil fertility parameters on modern rice productivity utilizing a stochastic production frontier approach on a survey data from 21 villages in three agro-ecological regions of Bangladesh. Detailed crop input-output data were collected from 380 paddy rice (Oryza sativa) farmers. The soil fertility status in each region was determined by analysis of soil organic carbon, available nitrogen, phosphorus and potassium concentration. Results reveal that in addition to key production inputs, soil fertility also significantly affects the parameters of the production function. Available soil potassium and soil nitrogen significantly increase rice productivity whereas available soil phosphorus has an opposite effect. The soil organic carbon content also has a desirable positive effect but the influence is not significant. The mean technical efficiency of these modern rice farmers is estimated at 73%, implying that 37% [(100-73)/73] of the production can be increased by eliminating technical inefficiency alone. Our results indicate that in policy terms technological initiatives should be targeted at measures to identify areas of lower soil fertility so that inherent soil-based productivity restrictions can be minimized. In part this will be facilitated by the transfer of indigenous knowledge from farmers in higher productivity areas, thus increasing rice production. Also, investment in rural infrastructure is suggested to improve technical efficiency.

---

* Address for Correspondence: Dr Sanzidur Rahman ; Senior Lecturer in Rural Development; School of Geography ; University of Plymouth; Drake Circus; Plymouth, PL4 8AA; England, UK; Phone: +44-1752-585911; Fax: +44-1752-585998 ; E-mail: srahman@plymouth.ac.uk. The present chapter draws heavily on materials from Rahman, S. and Parkinson, R.J. (2007), "Productivity and soil fertility relationships in rice production systems, Bangladesh", *Agricultural Systems*, Vol. 92(1-3): 318-333.

JEL classification: O33; Q18; C21

Keywords: Soil fertility, Rice production, Stochastic production frontier, Technical efficiency, Bangladesh.

1. INTRODUCTION

Land is the most important natural resource that provides livelihood for the majority of people in Bangladesh. Agriculture accounts for 23.5% of national income and employs 62% of the labor force (MoA, 2008). The dominant sector is the field crop agriculture accounting for more than 60% of agricultural value added. Among the field crops, rice is the major staple crop, occupying 70% of the cropped area (BBS, 2001). Historically, being a food deficit country with an extremely unfavourable land-person ratio of only 0.06 ha, Bangladesh has pursued a policy of rapid technological progress in agriculture by promoting diffusion of a rice-based 'Green Revolution (GR)' technology package. As a result, land-use intensity increased sharply to 175% in 1999 from its initial level of 146% in 1969 (Rahman and Thapa, 1999; and BBS, 2001) with corresponding increase in input use. For example, use of chemical fertilizers increased six times during 1968–94 and use of pesticides increased three-fold in just one decade during 1982–92 (Rahman and Thapa, 1999). Consequently, total rice output grew at an annual rate of 2.2 % during 1965-87 and then declined to half the previous rate at 1.1% during 1988-97 (Otsuka, 2000). In fact, the yield rate of modern rice steadily declined from 3.6 mt in 1969 to 2.4 mt in 1994 with an estimated annual rate of decline of 1.2% (Rahman, 2002), thereby confirming that productivity from GR technology is falling. It is believed that more than 65% of the total agricultural land is suffering from declining soil fertility and about 85% of net area suitable for cultivation has an organic matter below the minimum requirement (TFR, 1991). Soil analysis of 460 samples from 43 profiles from the same locations between 1967 and 1995 revealed a decline in fertility (Ali et al., 1997) although this decline has not been explicitly linked to GR technology. Baanante et al., (1993) noted that the present level of food crop production in Bangladesh takes up an estimated 0.93 million tons of nutrients (N, P, K and S) from the soil annually. Furthermore, MoA (2008) reported that 11 out of a total of 30 agro-ecological zones of Bangladesh have lost soil fertility between 10–70% due to intensified crop cultivation over a 30 year period from 1968–1998 (table 1). Widespread adoption of GR technology was identified as a cause of significant soil degradation and declining crop yields in India (Singh, 2000; Yadav et al., 2000). Pimentel (1996) indicated that extensive use of fertilizers and pesticides to support the GR has caused serious public health and environmental damages worldwide, particularly in developing countries. Furthermore, it has been noted that continued, intensive production of rice has lead to yield reductions in some countries in Asia, explained in part by soil nutrient exhaustion (Doberman et al., 2002).

Table 1. Losses of fertility of soil by intensified crop cultivation, 1967-68 to 1997-98

Agro Ecological zone (Number)	Types of land	Increase of cropping intensity (%)	Losses of Soil fertility (%)
Old Himalayann Piedmont Plain (1)	HL	100	25-45
Tista Floodplain (2)	HL	100	10-35
Tista Meander Floodplain (3)	HL	100	10-40
Old Brahmaputra Floodplain (9)	MHL	100	25-65
High Ganges River Floodplain (11)	HL	100	20-45
Middle Meghna River Floodplain (16)	MLL	100	15-40
Surma Kushiyara River Floodplain (20)	MLL	100	20-40
North Eastern Peidmont Plain (22)	HL	100	20-70
Chittagong Coastal Plain (23)	HL	100	10-30
Barind Tract (26)	HL	100	30-60
Madhupur Tract (28)	HL	100	40-65

Note: HL = High land, MHL = Medium high land, MLL = Medium low land.

The land type classification in Bangladesh is based on flooding depth. HL = no flooding, MHL = flooding depth of 0.01 – 0.90 m, MLL = flooding depth of 0.91 – 1.83 m, LL (Low land) = flooding depth of 1.83 – 3.05 m, VLL (Very low land) = flooding depth >3.05 m (Source: Land Resources Information database, Bangladesh Agricultural Research Council).

Source: Committee report for losses of soil fertility, 2004, BARI. (MoA, 2007, table 4.04).

There is a large body of literature on GR covering several dimensions of this complex technology package, such as productivity, growth, employment and equity (e.g., Das, 2002; Rahman, 1999; Freebairn, 1995; Hazell and Ramasamy, 1991), but the interactions between soil fertility status and farmers' indigenous knowledge have been less well studied. Payton et al. (2003) identified that Bangladeshi farmers possess an intimate and sophisticated knowledge of soil properties and management problems. Ali (2003) concluded that "despite their lack of knowledge of soil genesis, soil morphology and soil chemistry, farmers were able to qualitatively identify major typology, properties, and productivity constraints of topsoil" (pp. 333). This paper, therefore, contributes to the existing literature by explicitly examining the impact of soil fertility variables on the productivity performance of modern rice at farm-level in Bangladesh, a country that is particularly vulnerable in terms of food security.

2. Methodology

2.1. Analytical Framework

The stochastic production frontier approach, developed by Aigner et al., (1977), is utilized in this study. We extend the framework and include variables representing soil nutrient contents in addition to physical inputs to explain productivity performance. The stochastic production frontier for the *i*th farmer is written as:

$$Y_i = f(X_i, S_i) - u_i + v_i, \qquad (1)$$

where Y_i is the output, X_i is the vector of physical inputs, S_i is the vector of soil fertility variables, v_i is assumed to be independently and identically distributed $N(0,\sigma^2_v)$ two sided random error, independent of the u_i; and the u_i is a non-negative random variable $(u_i \geq 0)$, associated with inefficiency in production which is assumed to be independently distributed as truncation at zero of the normal distribution with mean $-Z_i\delta$, and variance σ_u^2 ($|N(-Z_i\delta, \sigma^2_u|$), where Z_i are the correlates of inefficiencies on farm i. In this formulation, output is assumed to be strictly monotonically increasing in both physical inputs as well as soil fertility variables.

In determining the predictors of production efficiency, we use the single stage approach proposed by Battese and Coelli (1995) wherein the technical inefficiency parameter is related to a vector of farm-specific managerial and household characteristics subject to statistical error, such that:

$$u_i = Z_i\delta + \zeta_i \geq 0, \quad (2)$$

where, Z_i are the farm-specific managerial and household characteristics and the error ζ_i is distributed as $\zeta_i \sim N(0, \sigma_\zeta^2)$. Since $u_i \geq 0, \zeta_i \geq -Z_i\delta$, so that the distribution of ζ_i is truncated from below at the variable truncation point, $-Z_i\delta$.

The production efficiency of farm i in the context of the stochastic frontier production function is defined as:

$$EFF_i = E[\exp(-u_i) \mid \xi_i] = E[\exp(-\delta_0 - \sum Z_i\delta \mid \xi_i) \quad (3)$$

where E is the expectation operator. This is achieved by obtaining the expressions for the conditional expectation u_i upon the observed value of ξ_i, where $\xi_i = v_i - u_i$. The method of maximum likelihood is used to estimate the unknown parameters, with the stochastic frontier and the inefficiency effects functions estimated simultaneously. The likelihood function is expressed in term of the variance parameters, $\sigma^2 = \sigma_v^2 + \sigma_u^2$ and $\gamma = \sigma_u^2 / \sigma^2$ (Battese and Coelli, 1995).

2.2. Location and Cropping Intensity in the Study Regions

Three of the intensive rice producing areas of Bangladesh, Jamalpur, Jessore and Comilla, were selected for this study. Jamalpur is located within Jamalpur Sadar Thana (central administrative sub-district), in the south-eastern part of Jamalpur district. The study area is 180 km northwest from Dhaka. Jessore is located in Manirampur Thana in the southern part of Jessore district, 290 km southwest from Dhaka. Comilla is located in Matlab Thana in the south-eastern part of Chandpur district, 120 km southeast from Dhaka.

An estimated 75% of the gross cropped area was under modern varieties of rice and wheat and 62% was under irrigation in the study regions during the crop year 1996. The cropping intensity (defined as the ratio of gross cropped area to net sown area multiplied by 100) of the sample farms is estimated at 172.8 (183.3 in Jamalpur, 178.2 in Jessore and 148.2

in Comilla region), which is remarkably close to the national estimate of 173.2 for the year 1995/96 (BBS, 2001).

2.3. Data and Variables

The study is based on farm-level data for crop year 1996 collected from three agro-ecological regions of Bangladesh. The survey was conducted from February to April 1997. The specific selected regions were Jamalpur (representing wet agro-ecology), Jessore (representing dry agro-ecology), and Comilla (representing both wet agro-ecology and an agriculturally developed area). A multistage random sampling technique was employed to locate the districts, then the *Thana* (sub districts), and then the villages in each of the three sub districts and finally the sample households. A total of 380 households from 21 villages (174 households from eight villages of Jamalpur Sadar *Thana*, 100 households from six villages of Manirampur *Thana* and 106 households from seven villages of Matlab *Thana*) form the sample for the study. Detailed input-output data were collected for modern varieties of rice produced in a crop year. The dataset also includes information on level of soil fertility determined from soil samples collected from representative locations in the study villages.

2.4. Soil Sampling and Analysis

Data on physical and chemical properties of soils from the selected farmers' fields were collected to evaluate the fertility status of the soil and to examine inter-regional differences between the study areas. Soils were mapped in the three study areas at series level. Soil series were distinguished following a process of detailed chemical and physical characterisation using standard procedures (SRDI, 1991), based on assessment of inherent variability. SRDI (1991) employed ranges and thresholds for key soil parameters (texture, colour, structure, pH, organic carbon, available phosphorus, potassium and iron) to define and map separate soil series. Therefore soil series as mapped were used to select five distinct sampling locations in each region, giving a total of 15 composite soil samples in total, representing 15 different soil series. Soils were collected from random locations within rice-plots within the survey households. Soil sub-samples were collected from the 0–200 mm cultivated horizon at each of 3–5 random locations within a selected plot, and were then thoroughly mixed to give a composite sample. Samples were air dried prior to analysis.

As a part of a wider project (Rahman, 1998), soil samples were analyzed for (1) soil organic carbon content, (2) available potassium, (3) available phosphorus, (4) available nitrogen, (5) available sulphur, (6) available zinc, (7) soil texture, and (8) soil pH. In this paper, data are presented for the key soil chemical parameters that are directly modified by routine fertilizer practice and are less subject to inter-season and spatial variability, viz. soil organic carbon content, available phosphorus, available potassium, and available nitrogen. Soil organic carbon (SOC) content was measured using the Walkley-Black rapid titration method. Available phosphorus (P) was extracted using Truog's extraction method and determined colorimetrically by spectrophotometer. Available potassium (K) was extracted by neutral normal ammonium acetate solution and determined by Gallenkamp flame photometer. Available (mineral) nitrogen (N) was determined using the micro-Kjeldahl method. Full methodological details of all soil analyses are given in PCARRD (1980). Soil organic carbon, available N, P and K concentrations were converted into topsoil mass per unit area (kg ha^{-1}) assuming a soil bulk density of 1.0 Mg m^{-3} for the cultivated soil horizon (0.00–0.20 m).

2.5. Soil Fertility Variables

In order to create the soil fertility variables for each of the sampled households, we extrapolated our soil sample results taken from representative farm-plots to farm households whose plots belong to the same village and fall within the same soil series. The justification for adopting this approach is two-fold. First, in preparing the Land and Soil Resource Use Guideline at *Thana* (sub-district) level, which includes a soil fertility map based on representative soil sampling, the Soil Development Research Institute utilized the soil series classification system (SRDI, 1991). They noted that a particular soil series is named/determined according to a number of characteristics, e.g., texture, structure, colour, organic matter content, and soil pH. Also, in general, chemical properties of soils within a given soil series were observed to be similar, as soils mapped within each series was formed from the same parent material. Therefore it was assumed that each soil series possessed similar properties. Hence, one can expect similar results from other soil samples collected from the same soil series. Although it should be noted that there are obvious limitations to this extrapolation, given that many soil properties are dynamic and depend not only on parent material but also on recent management history, which was found to be similar across surveyed farmers. As such, one can use replicate information on soil properties collected from one location of a given soil series to other locations falling within the same soil series simply by referring to the soil fertility map (SRDI, 1991). Second is the nature of the paddy fields from where these samples were taken. We took soil samples from rice plots of surveyed households, which were located within a large continuous block of land area designated as paddy fields where most of the farmers of that village have rice plots. Furthermore, cross-referencing of these blocks of paddy fields with the soil fertility map revealed that they all fell within the same soil series. Therefore, extrapolation of soil fertility parameters to surveyed households with rice plots falling within the same soil series was not expected to pose any significant limitation to the interpretation of soil analysis data (Rahman and Parkinson, 2007).

2.6. The Empirical Model

The production structure of modern rice farmers in Bangladesh is specified using a single output multiple input stochastic production frontier. The general form of the extended flexible translog stochastic production frontier for the ith farm is defined as:

$$\begin{aligned}\ln Y_i = {} & \alpha_0 + \sum_{j=1}^{5} \alpha_j \ln X_{ij} + \tfrac{1}{2}\sum_{j=1}^{5}\sum_{k=1}^{5} \beta_{jk} \ln X_{ij} \ln X_{ik} \\ & + \sum_{m=1}^{4} \eta_m \ln W_{im} + \tfrac{1}{2}\sum_{m=1}^{4}\sum_{n=1}^{4} \kappa_{mn} \ln W_{im} \ln W_{in} + \tfrac{1}{2}\sum_{m=1}^{4}\sum_{j=1}^{5} \tau_{mj} \ln W_{im} \ln X_{ij} + v_i - u_i\end{aligned} \quad (4)$$

and

$$u_i = \delta_0 + \sum_{d=1}^{7} \delta_d Z_{id} + \zeta_i \tag{4a}$$

where the dependent variable Y is yield of rice output produced per ha; X's are the inputs used per ha; W's are the soil variables; and Z's are the variables representing farms' socio-economic circumstances, where

X_j = quantity of input used per ha of land cultivated,
j = 1, fertilizers (kg ha^{-1})
= 2, labour (person days ha^{-1})
= 3, animal power services (animal pair-days ha^{-1})
= 4, pesticides (taka ha^{-1})
= 5, proportion of irrigation (%)
W_m = soil fertility parameters per ha of land,
m = 1, soil available phosphorus (kg ha^{-1})
= 2, soil organic carbon content (kg ha^{-1})
= 3, soil available potassium (kg ha^{-1})
= 4, soil available nitrogen (kg ha^{-1})
ε = random error
ln = natural logarithm

The variables in the inefficiency effects function are

Z_d = socio-economic factors,
= 1, tenurial status (1 = if owner operator, 0 otherwise)
= 2, farming experience (years)
= 3, education of farmer (completed year of schooling)
= 4, subsistence pressure (number of persons per household)
= 5, share of non-agricultural income (proportion)
= 6, extension contact (1 = if had contact in the past one year, 0 otherwise)
=7, index of underdevelopment of infrastructure[1] (number)
α_0, α_j, β_{jk}, η_m, κ_{mn}, τ_{mj} and δ_{id} are the parameters to be estimated.

Fertilizer, labour and animal power, are the three major inputs that are essential in producing any crop and contribute significantly to total cost of production (Rahman, 1999). Owing to diffusion of GR, pesticide also became an integral part of the system, although past studies consistently omitted this essential input, except for some in recent years, such as Tzouvelekas et al. (2001) and Wadud and White (2000). Lack of access to irrigation has been identified as one of the principal reasons for stagnation in GR diffusion in Bangladesh

[1] A composite 'index of underdevelopment of infrastructure' was constructed using the cost of access approach. A total of 13 elements are considered for its construction. These are primary market, secondary market, storage facility, rice mill, paved road, bus stop, bank, union office, agricultural extension office, high school, college, thana (sub-district) headquarters, and post office (see Ahmed and Hossain, 1990 for construction details).

(Rahman and Thapa, 1999). All these variables are expected to have significant positive association with yield of rice output. Inclusion of soil-related ‘state-of-nature’ variables is a rather uncommon practice in farm economic analysis (Sidhu and Baanante, 1981). Moreover, given the emerging concerns regarding sustainability of food production, declining soil fertility and other environmental problems arising from GR adoption (e.g., Shiva, 1991; Pimentel, 1996; Rahman and Thapa, 1999; and Singh, 2000), inclusion of these variables in economic decision making is becoming more and more important. The key soil fertility variables used in this analysis were soil organic carbon and available N, P and K, all of which are core components of soil fertility (Parkinson, 2003). Soil organic carbon (SOC) is an important soil fertility indicator because it has a major influence on biophysical and biochemical soil function. Soil pH was not included because the range of pH values observed was restricted, and fell within the optimum range suitable for rice production environment (SRDI, 1991). Our *a priori* expectation is that all these soil nutrients will have a positive influence on rice yield.

3. Results

Summary statistics of the variables used in the stochastic production frontier model are presented in table 2. Soil fertility variables included in the analysis were SOC, indicative of inherent soil physical, chemical and biological fertility, and available soil N, P and K, the three macronutrients that are yield limiting in rice production systems (Wijnhoud *et al.*, 2003).

The Maximum Likelihood Estimation (MLE) procedure is used to estimate the parameters of the stochastic production frontier and inefficiency effect models jointly in a single stage[2] using STATA Version 8 (Stata Corp, 2003) and the results are presented in table 3. A test of hypothesis on the choice of functional form (Cobb-Douglas vs. translog, i.e., H_0: $\beta_{jk} = 0$ for all jk) confirms that the choice of translog production function is a better representation of the production structure (LR χ^2 with 45 df = 96.39, $p<0.000$).

All production inputs including soil fertility variables have the expected positive sign except soil available P. Among the production inputs, labor, pesticides and irrigation are the significant determinants of rice productivity, as expected since modern rice is labor intensive and need supplementary irrigation as well as pesticides. The output elasticity of labor is highest estimated at 0.15 indicating that a 1% increase in labor use will increase rice yield by 0.15%[3].

[2] The single-stage approach is considered superior to the conventionally used two-stage approach wherein the first stage involves estimation of the stochastic production frontier and the prediction of inefficiency effects under the assumption that these inefficiency effects are identically distributed with one-sided error terms. The second stage involves the specification of a regression model for predicted inefficiency effects, which contradicts the assumption of an identically distributed one-sided error term in the stochastic frontier (Battese and Coelli, 1995).

Table 2. Description, measure and summary statistics of the variables

Name	Description	Measurement	Mean	Standard deviation
Production variables				
Y	Yield of modern rice output	kg ha^{-1}	4218.80	1110.09
Land	Land area under modern rice	ha	0.73	0.79
F	Quantity of fertilizers	kg ha^{-1}	251.78	73.18
L	Quantity of labor	person-days ha^{-1}	105.31	47.90
A	Quantity of animal power services	animal pair-days ha^{-1}	27.48	8.78
I	Quantity of pesticides	taka ha^{-1}	232.61	335.84
W	Irrigation	proportion of land area under rice that is irrigated	0.65	0.28
P	Soil available phosphorus	kg ha^{-1}	44.63	15.22
O	Soil organic carbon	kg ha^{-1}	40.28	31.34
K	Soil available potassium	kg ha^{-1}	68.58	31.63
N	Soil available nitrogen	kg ha^{-1}	38.11	11.98
Farm-specific variables				
Z_1	Tenurial status	1 if owner operator, 0 otherwise	0.58	0.49
Z_2	Farming experience	years	25.26	14.20
Z_3	Education of the farmer	completed years of schooling	3.69	4.28
Z_4	Subsistence pressure	persons per household	5.98	2.52
Z_5	Share of non-agricultural income	proportion of total income	0.19	0.29
Z_6	Extension contact	1 if had contact, 0 otherwise	0.13	0.33
Z_7	Infrastructure index	number	34.25	14.88
	Number of observations		380	

Note: [a] = Exchange rate of USD 1.00 = Taka 42.70 in 1996 (BBS, 2001).

Table 3 also shows the influence of soil fertility variables on rice productivity. SOC is a commonly accepted indicator of soil fertility, there being a positive relationship between organic carbon content and productivity in most cropping systems (Parkinson, 2003). Although we see positive influence of SOC on rice output, the coefficient is not significantly different from zero. On the other hand, an increase in available N and K significantly increases rice yield, confirming well known relationships between soil fertility and crop productivity (Yadav, 2003), although the magnitude of influence is much higher for available K. The output elasticity measure indicates that a 1% increase in soil available K will increase rice yield by almost 13%. It is, however, surprising to see a significant negative impact of soil available P on rice yield. This observation is counterintuitive, and needs further investigation. Overall, our results confirm that for these modern rice farmers, high inherent soil fertility leads to greater yields.

[3] All the resource input variables were mean-corrected prior to estimation. Therefore, the coefficients on the first order term can be read directly as elasticities.

Table 3. Stochastic production frontier estimates of modern rice production

Variables	Coefficient	t-ratio
Production function		
Constant	-30.5024	-0.41
ln Fertilizer (F)	0.0638	1.41
ln Labor (L)	0.1447	4.23***
ln Animal power (A)	0.0338	0.59
ln Pesticides (I)	0.0193	2.18**
ln Irrigation (W)	0.0548	1.83*
ln Soil available phosphorus (P)	-12.0616	-2.06**
ln Soil organic carbon (O)	2.0393	1.23
ln Soil available potassium (K)	13.5566	2.62***
ln Soil available nitrogen (N)	6.4912	1.70*
ln $(\text{Fertilizer})^2$	0.0128	0.14
ln $(\text{Labor})^2$	0.4218	4.33***
ln $(\text{Animal power})^2$	-0.2860	-0.97
ln $(\text{Pesticides})^2$	0.0034	0.49
ln $(\text{Irrigation})^2$	0.0110	1.76*
ln $(\text{Soil phosphorus})^2$	183.1650	2.07**
ln $(\text{Soil organic carbon})^2$	-12.8693	-2.07**
ln $(\text{Soil potassium})^2$	11.2065	2.47***
ln $(\text{Soil nitrogen})^2$	-45.5224	-2.47***
ln F x ln L	0.0264	0.29
ln F x ln A	0.0341	0.21
ln F x ln I	-0.0123	-0.81
ln F x ln W	-0.0143	-0.93
ln F x ln P	0.1814	0.92
ln F x ln O	0.0221	0.27
ln F x ln K	-0.1320	-0.99
ln F x ln N	0.0328	0.22
ln L x ln A	0.0611	0.40
ln L x ln I	-0.0121	-0.85
ln L x ln W	-0.0101	-0.39
ln L x ln P	-0.1217	-0.81
ln L x ln O	0.0641	0.93
ln L x ln K	0.0791	0.71
ln L x ln N	-0.0253	-0.20
ln A x ln I	-0.0252	-0.99
ln A x ln W	-0.0136	-0.51
ln A x ln P	0.1876	0.69
ln A x ln O	-0.0743	-0.61
ln A x ln K	0.0047	0.03
ln A x ln N	0.3354	1.61
ln I x ln W	0.0019	0.65
ln I x ln P	-0.0589	-2.05**

ln I x ln O	-0.0082	-0.62
ln I x ln K	-0.0231	-1.48
ln I x ln N	0.0082	0.41
ln W x ln P	0.0245	0.5
ln W x ln O	-0.0454	-2.20**
ln W x ln K	0.1113	1.74*
ln W x ln N	-0.0401	-1.08
ln P x ln O	58.2921	2.27**
ln P x ln K	-22.2185	-2.49***
ln P x ln N	29.8759	1.25
ln O x ln K	-5.0747	-2.40***
Variables	Coefficient	t-ratio
ln O x ln N	37.6216	2.34**
ln K x ln N	-55.6834	-2.55***
Model diagnostics		
$\sigma^2 = \sigma_u^2 + \sigma_v^2$	0.0434	13.78***
$\gamma = \sigma_u^2/(\sigma_u^2 + \sigma_v^2)$	0.5900	2.04**
Log likelihood		
Technical inefficiency predictors		
Constant	0.0091	0.10
Tenurial status	-0.0093	-0.38
Farming experience	0.0022	2.37***
Education of the farmer	0.0057	1.73*
Subsistence pressure	0.0058	1.15
Share of non-agricultural income	0.0663	1.54
Extension contact	-0.0390	-1.05
Infrastructure index	0.0057	4.12***
Number of observations	380	

Note: *** significant at 1 % level ($p<0.01$)
** significant at 5 % level ($p<0.05$)
* significant at 10 % level ($p<0.10$).

3.1. Determinants of Technical Inefficiency

Prior to the discussion of the predictors of technical inefficiency, we briefly report the technical efficiency scores (table 4). It is clear from table 4 that high level of inefficiency exists among these modern rice farmers. The mean technical efficiency of these rice farmers is estimated at 73% implying that 37% [(100-73)/73] of rice yield can be increased by eliminating technical inefficiency alone. The estimate, however, falls within the common range of 69 – 82% reported in the literature[4] (e.g., Bravo-Ureta et al., 2007).

[4] Bravo-Ureta et al., (2007), using a meta-analysis of 167 efficiency studies conducted worldwide, arrived at this range of technical efficiency levels. For Asia in particular, the mean technical efficiency is 74.9% when stochastic production frontier approach is used, which is remarkably close to our estimates.

Table 4. Technical efficiency in rice production

Variables	Efficiency scores
Efficiency levels	
Up to 60 %	4.21
61 – 70 %	26.58
71 – 80 %	55.79
81 – 90 %	13.42
90 and above	--
Mean efficiency level	0.73
Standard deviation	0.62
Minimum	0.53
Maximum	0.87
Number of observations	380

The lower panel of table 3 presents the results of the predictors of technical inefficiency. The null hypotheses with regard to the existence of inefficiency and validity of the specified predictors of inefficiency were tested (i.e., H_0: $\mu = \gamma = 0$) and strongly rejected at the 1% level of significance (LR χ^2 with 7 df = 29.11, p<0.000).

Farming experience has a significant negative impact on technical efficiency[5], implying that older farmers are more inefficient, which is consistent with the existing literature[6] (e.g., Seyoum et al., 1998; Llewelyn and Williams, 1996; Dhungana et al., 2004). Farmers' education also seems to have a significant negative influence on technical efficiency. The impact of education on technical efficiency is mixed in the literature. Although many claim positive impact on technical efficiency (e.g., Seyoum et al., 1998; Ali and Flinn, 1989; Young and Deng, 1999), some find no impact (e.g., Battese and Coelli, 1995; Llewellyn and Williams, 1996), and few also claim negative impact (e.g., Hasnah et al., 2004). The latter seems to corroborate with our finding. The implication is that educated farmers substitute their time away from farm production and are, therefore, relatively inefficient compared to their peers. Farmers located in regions endowed with better infrastructure are more technically efficient, as expected[7], which is consistent with the existing literature (e.g., Ali and Flinn, 1989; Wang et al., 1996; Wadud and White, 2000). Infrastructure development opens up opportunities for marketing, storage and resource supplies, which in turn complements production performance by enabling timely purchase and use of required inputs. For example, Ahmed and Hossain (1990) concluded that farms in villages with relatively developed infrastructure use relatively greater amounts of fertilizer and market a higher percentage of their agricultural products in Bangladesh.

[5] A positive sign on the coefficient indicates negative impact on technical efficiency.

[6] Use of farming experience as a variable explaining technical efficiency is not very common in the literature. The most common measure is the age of the farmer, but the conclusion remains the same.

[7] This index is constructed as the "underdevelopment of infrastructure". Therefore, a positive sign on the coefficient of this variable implies a positive impact on technical efficiency.

4. CONCLUSION

The inclusion of soil fertility variables in productivity and efficiency analysis is uncommon, although in reality farmers' production performance is highly likely to be influenced by both economic as well as bio-physico-chemical factors. The present study attempted to integrate these two strands of scientific enquiry when analyzing farmers' production performance. Therefore, we incorporated four key soil fertility variables and examined their influence on rice productivity while explicitly controlling for farmers' production inputs. Our results have demonstrated a clear relationship between soil fertility (a combined bio-physico-chemical factor) and rice productivity in the Jamalpur, Jessore and Comilla regions of Bangladesh. The yield of rice is significantly higher in fertile regions, as expected. We have shown that there is a close relationship between rice yield and disaggregated indicators of soil fertility. This observation emphasises the importance of developing regional agricultural policy approaches that allow the transfer of indigenous knowledge, as farmers do not carry out either detailed nutrient budgets or soil nutrient analysis on a routine basis.

The overarching policy implication of this study is that enhancement of soil fertility exerts a positive impact on rice yield, thereby raising farmers' income. Enhancement of soil fertility status through more efficient nutrient utilisation will be beneficial on both economic and bio-physico-chemical grounds. Therefore, government policies should be geared towards devising an effective strategy that promotes soil conservation measures to ensure and sustain future productivity potential of these soils. The farmers surveyed here, operating on a range of soils of contrasting fertility status, were able to adjust input use effectively based on soil observation and indigenous knowledge (Payton et al., 2003). However, there are other agro-ecological zones in Bangladesh where lower natural soil fertility and lack of understanding of soil fertility limits crop production severely (Rahman and Parkinson, 2007). In these situations, the transfer of inherent soil fertility knowledge acquisition skills is a necessary prerequisite for raising farm productivity in these less fertile areas, and will result in a more efficient utilisation of nutrients. It is recommended that knowledge transfer of agro-economic advice at village level, emphasising soil nutrient budgeting, be implemented in order to allow increased nutrient utilisation efficiency in regions where soil fertility is depleted. Challenges remain, as the transfer of technological and intuitive knowledge is restricted by the lack of routine soil and crop monitoring procedures at farm level. There is a requirement to focus agronomic advice and support at the local level, and hence facilitate the transfer of indigenous knowledge from farmers in higher productivity areas. Another area of intervention to improve technical efficiency in rice farming is to invest in rural infrastructure development.

REFERENCES

Ahmed, R., Hossain, M. 1990. Developmental impact of rural infrastructure in Bangladesh. IFPRI Research Report 83. International Food Policy Research Institute, Washington, D.C.

Aigner, D.J., Lovell, C.A.K., Schmidt, P. 1977. Formulation and estimation of stochastic frontier production function models. *Journal of Econometrics*, 6: 21 – 37.

Ali, A.M.S. 2003. "Farmers' knowledge of soils and the sustainability of agriculture in a saline water ecosystem in South-western Bangladesh". *Geoderma*, 111: 333 – 353.

Ali, M., Flinn, J.C., 1989, Profit efficiency among Basmati rice producers in Pakistan Punjab. *American Journal of Agricultural Economics*, 71, 303-310.

Ali, M.M., Shaheed, S.M., Kubota, D., Masunaga, T., Wakatsuki, T. 1997. "Soil degradation during the period 1967 – 1995 in Bangladesh, 2: selected chemical characters". *Soil Science and Plant Nutrition*, 43: 879 – 890.

Baanante, C.A., Henao, J., Wan, X. 1993. *Fertilizer Subsidy Removal in Bangladesh: An Assessment of the Impact of Fertilizer Use, Crop Yields, and Profits of Farmers*. International Fertilizer Development Centre, Muscle Shoals, Alabama.

Battese, G.E., Coelli, T.J. 1995. A model for technical inefficiency effects in a stochastic frontier production function for panel data. *Empirical Economics*, 20, 325-332.

BBS. 2001. *Yearbook of Agricultural Statistics of Bangladesh, 1999*. Bangladesh Bureau of Statistics, Dhaka.

Bravo-Ureta, B.E., Solis, D., Lopez, V.H.M., Maripani, J.F., Thiam, A., Rivas, T. 2007. Technical efficiency in farming: a meta regression analysis. *Journal of Productivity Analysis,* 27:57-72.

Das, R.J. 2002. "The green revolution and poverty: a theoretical and empirical examination of the relation between technology and society". *Geoforum*, 33: 55 – 72.

Dobermann, A., Witt, C., Dawe, D., 2002. "Site-specific nutrient management for intensive rice cropping systems in Asia". *Field Crops Research*, 74: 37-66.

Dhungana, B.R., Nuthall, P.L., Nartea, G.V. 2004. "Measuring the economic inefficiency of Nepalese rice farms using data envelopment analysis". *The Australian Journal of Agricultural and Resource Economics*, 48, 347-369.

Freebairn, D.K. 1995. "Did the Green Revolution concentrate incomes? A quantitative study of research reports". *World Development*, 23: 265 – 279.

Hasnah, Fleming, E.M., Coelli, T., 2004, Assessing the performance of a nucleus estate and smallholder scheme for oil palm production in West Sumatra. *Agricultural Systems*, 79, 17-30.

Hazell, P.B.R., Ramasamy, C. 1991. *The Green Revolution Reconsidered.* Baltimore: Johns Hopkins University Press.

Llewelyn, R.V., Williams, J.R., 1996. Nonparametric analysis of technical, pure technical, and scale efficiencies for food crop production in East java, Indonesia. Agricultural Economics, 15: 113-126.

MoA, 2008. Handbook of Agricultural Statistics, December 2007. Ministry of Agriculture, http://www.moa.gov.bd/statistics/statistics.htm

Otsuka, K. 2000. "Role of agricultural research in poverty reduction: lessons from the Asian experience". *Food Policy*, 25: 447 – 462.

Parkinson, R.J. 2003. Soil Management and Crop Nutrition, in Soffe, R. J. (Ed). The Agricultural Notebook 20th Edition. Blackwell Scientific Press, p 3-62

Payton, R.W., Barr, J.J.F., Martin, A., Sillitoe, P., Deckers, J.F., Gowing, J.W., Hatibu, N., Naseem, S.B., Tenywa, M., Zuberi, M.I. 2003. "Contrasting approaches to integrating indigenous knowledge about soils and scientific soil survey in East Africa and Bangladesh". *Geoderma*, 111: 355 – 386.

PCARRD. 1980. Manual: standard methods for analysis of soil, plant, tissue, water and fertilizer. Philippines Council for Agriculture and Forestry Research, Los Banos, Philippines.

Pimentel, D. 1996. "Green Revolution and chemical hazards". *The Science of the Total Environment*, 188 (Supplement 1): S86 – S98.

Rahman, S. 1998. Socio-economic and environmental impacts of technological change in Bangladesh agriculture. Unpublished Ph.D. Dissertation. Asian Institute of Technology (AIT), Bangkok, Thailand.

Rahman, S. 1999. "Impact of technological change on income distribution and poverty in Bangladesh agriculture: an empirical analysis". *Journal of International Development*, 11: 935 – 955.

Rahman, S. 2002. "Technological change and food production sustainability in Bangladesh agriculture". *Asian Profile*. 30: 233 – 246.

Rahman, S., Thapa, G.B. 1999. "Environmental impacts of technological change in Bangladesh agriculture: farmers' perceptions and empirical evidence". *Outlook on Agriculture*. 28: 233 – 238.

Rahman, S. Parkinson, R.J. 2007. "Productivity and soil fertility relationships in rice production systems, Bangladesh". *Agricultural Systems*. 92: 318 – 333.

Seyoum, E.T., Battese, G.E., Fleming, E.M., 1998, "Technical efficiency and productivity of maize producers in eastern Ethiopia: a study of farmers within and outside the Sasakawa Global 2000 project". *Agricultural Economics*, 19, 341-348.

Shiva, V. 1991. *The Violence of the Green Revolution: Third World Agriculture, Ecology and Politics*. Zed Books, London.

Sidhu, S.S., Baanante, C.A. 1981. "Estimating farm-level input demand and wheat supply in the Indian Punjab using a translog profit function". *American Journal of Agricultural Economics*, 63: 237 – 246.

Singh, R.B. 2000. "Environmental consequences of agricultural development: a case study from the Green Revolution state of Haryana, India". *Agriculture, Ecosystem and Environment*. 82: 97 – 103.

SRDI. 1991. *Soil Guides for Crop Production, 1991 (in Bangla.)* Soil Resource Development Institute, Dhaka.

STATA Corp, 2003. *STATA Version 8*. Stata Press Publications, College Station, Texas, USA.

TFR, 1991. *Bangladesh Development Strategies: Environment and Development*. Vol. 4. Dhaka: University Press Ltd.

Tzouvelekas, V., Pantzios, C.J., Fotopoulos, C. 2001. "Technical efficiency of alternative farming systems: the case of Greek organic and conventional olive growing farms". *Food Policy*, 26: 549 – 569.

Wadud, A., White, B. 2000. "Farm household efficiency in Bangladesh: a comparison of stochastic frontier and DEA methods". *Applied Economics*, 32: 1665 – 1673.

Wang, J., Cramer, G.L., Wailes, E.J. 1996. "Production efficiency of Chinese agriculture: evidence from Rural Household Survey data". *Agricultural Economics*, 15, 17 – 28.

Wijnhoud, J.D., Konboon, Y., Lefroy, R.D.B. 2003. "Nutrient budgets: sustainability assessment of rainfed lowland rice-based systems in northeast Thailand". *Agriculture, Ecosystems and Environment*. 100: 119-127.

Yadav, R.L, Dwivedi, B.S. Prasad, K., Tomar, O.K. Shurpali, N.J. Pandey, P.S. 2000. "Yield trends, and changes in soil organic-C and available NPK in a long-term rice-wheat system under integrated use of manures and fertilizers". *Field Crops Research* 68, 219-246.

Yadav, R. L. 2003. "Assessing on-farm efficiency and economics of fertilizer N, P and K in rice wheat systems of India". *Field Crops Research.* 81, 39-51.

Young, D., Deng, H., 1999, "The effects of education in early-stage agriculture: some evidence from China". *Applied Economics*, 31, 1315-1323.

In: Soil Fertility
Editors: D. P. Lucero and J. E. Boggs

ISBN: 978-1-60741-466-71

Chapter 4

A COMPARISON BETWEEN SHORT-TERM EFFECTS OF SEWAGE SLUDGE COMPOST AND PHOSPHATE SALT FERTILIZATIONS ON MICROBIAL ACTIVITIES INVOLVED IN P TURN-OVER OF TWO CALCAREOUS AND SILICEOUS MEDITERRANEAN VINEYARD SOILS

Stéven Criquet* and Armelle Braud
Institut Méditerranéen d'Ecologie et de Paléoécologie, IMEP, UMR CNRS 6116,
Equipe Ecologie Microbienne, Service 452, Université Paul Cézanne,
Faculté des Sciences et Techniques de Saint-Jérôme, 13397 Marseille, Cedex 20, France

ABSTRACT

As part of a larger study, this chapter investigates the influence of different fertilization practices on microbial parameters involved in P turn-over of Mediterranean soils. More precisely, the aim of the present study was to compare the effects of both mineral and organic amendments on microbial patterns involved in P recycling of two Mediterranean soils. With this aim, the effects of both Na and K phosphate salts (supplied as NaH_2PO_4 and KH_2PO_4, respectively), as well as sewage sludge compost (SSC) on both acid (APH) and alkaline (BPH) phosphatases and phosphodiesterase (PDI) activities, BPH/APH ratio, densities of culturable heterotrophic bacteria and available P (P Olsen) were investigated in two Mediterranean vineyard soils originating from calcareous and siliceous mineral parent materials. All of the amendments were carried out in order to provide soil with a P total quantity equivalent to 0.5 g P_2O_5 per kg of soil, and variables were measured after 25 and 87 d of incubation under field conditions. Globally, results showed that neither phosphate salts nor SSC had significant effects on P mineralizing enzyme activities, nor on the densities of culturable heterotrophic bacteria. On the contrary, all of these amendments resulted in a significant increase in available P of the

* Corresponding author: Stéven Criquet; Institut Méditerranéen d'Ecologie et de Paléoécologie, IMEP, UMR CNRS 6116 ; Equipe Ecologie Microbienne, Service 452, Université Paul Cézanne; Faculté des Sciences et Techniques de Saint-Jérôme, 13397 Marseille, Cedex 20, France. Tel: +33-491-288530; fax: +33-491-288190; E-mail address: steven.criquet@univ-cezanne.fr

two soils considered. Results also showed a significant effect of the incubation time on several variables, especially APH and BPH activities, as well as available P. Indeed, between the two sampling dates most of these enzyme activities significantly decreased, whereas P availability was improved, SSC amended soils showing the more important variations in available P contents. This fertilization experiment also revealed that soil origin had only a significant effect on BPH activities, and that the nature of the cation (i.e., Na or K) within the different P salts used did not show any significant effect on the different variables investigated.

Keywords: available P, mineral P, P fertilization, phosphodiesterases, phosphomonoesterases, sewage sludge compost

1. INTRODUCTION

In most terrestrial ecosystems, P is known to be a limiting element for the development of organism populations. Even more, in an agronomic context, P is essential for agriculture productivity because of its crucial role in many metabolic processes involved in the growth of plants and soil microorganisms (Saggar et al., 1998). Thus, this element, as well as others such as C or N, is a key factor in soil quality and fertility problems (Hooda et al., 2001; Caravaca et al., 2002). To be assimilated by plants or microorganisms, P must be released in soil solution in its inorganic orthophosphate form, a process that can be driven by mineralization of organic P, desorption of available P from soil constituents or, to a lesser extent, by dissolution of P from bedrock (Santruckova et al., 2004). Soil P fertilization can be performed using phosphate salts or by a large diversity of organic fertilizers, the proportion of each depending mostly upon agriculture type and the agronomic practices of the country. Because of contamination risk of groundwater by phosphate salt fertilization, organic P sources are increasingly used to improve the available P pool of soils, as well as other nutrients involved in soil fertility (Melero et al., 2006). Among the organic fertilizers that can be used to fertilize soil, there is an increasing interest in the recycling of municipal organic wastes such as sewage sludge (SS) or sewage sludge compost (SSC) (Selivanovskaya and Latypova, 2006). Indeed, these kinds of wastes contain large amounts of nutrients that can significantly promote plant productivity but may also present some risks to the environment because of the very labile form of P and other nutrients, and because of the possibility of heavy metal accumulation in soil following repeated application of SS (Speir et al., 2004; Criquet et al., 2007). On this point, the European community has legislated the use of sewage sludge in agriculture by the Sewage Sludge Directive 86/278/EEC (1986), which "recommends regulating the utilization of sewage sludge in such a way as to prevent harmful effects on soil, vegetation, animals and man, thereby encouraging the correct use of sewage sludge". Therefore, in order to reduce the harmful effects of sludges without impeding their fertilization qualities, sludge composting (SSC) is a possibility that is increasingly used in agronomy. However, only a small P portion of SSC is in an inorganic orthophosphate form that can be directly assimilated by plants and microorganisms, while most of P is in an organic form requiring a preliminary enzymatic hydrolysis to release available PO_4^{3-} (Lima et al., 1996). In soils, acid and alkaline phosphomonoesterases, as well as phosphodiesterases, are the main enzymes involved in the release of PO_4^{3-} from phosphorylated organic matter

(Tabatabai, 1994; Criquet et al., 2004). These soil enzymes are mainly produced by plants and microorganisms, and they allow effective functioning of P turn-over within terrestrial ecosystems. Thus, knowledge of the factors that can influence their activity and their production are of particular importance in maintaining soil fertility and quality. For the same reason, the present study takes places within a larger one concerning the effects of several kinds of fertilizers on soil biological properties involved in P turn-over of different Mediterranean soils (Criquet et al., 2007; Criquet and Braud, 2008). With this aim we achieved a complete experimental design integrating the effects of several amendments constituted by three types of SS, one SSC and two types of mineral P-salts (NaH_2PO_4 and KH_2PO_4) on the microbial parameters involved in P recycling of four Mediterranean soils, i.e., two vineyard soils (from siliceous or calcareous origin), one degraded siliceous soil characterized by low level in soil organic matter and nutrients, and one calcareous soil commercialized by the Biotechna company. All amendment modalities, as well as controls, were applied to the different soils (7 x 4), with the exception of phosphate salts x Biotechna soils. In a first paper (Criquet et al., 2007), we focused on SS effects and showed that SS significantly increased phosphatase activities, microbial densities and available P of the two vineyards soils, but we also suggested that SS seemed to have only time-limited effects on soil fertility. This study also revealed that the chemical characteristics of sludges may show different effects on microbial parameters. Following this study, we aimed to compare the effects exerted by all of the different amendments described above on microbial patterns involved in P turn-over of the degraded soil. Results from our second study (Criquet and Braud, 2008) confirmed the first ones regarding SS effects. On the contrary, SSC and P-salt amendments did not affect microbial parameters in most cases, although they increased available P. Finally, with this third and last paper, we aimed to demonstrate whether response patterns previously described using a nonagricultural and degraded soil can be generalized to other soils—that is to say, if they are the same in the case of agro-systems constituted by both calcareous and siliceous vineyard soils subjected to SSC and mineral P fertilizations.

2. Materials and Methods

2.1. Origins and Sampling of the Soils

The soils were sampled in two vineyards at Château des Launes (La Garde Freinet, France) and the Domaine de Pourcieux (Pourcieux, France), both located in the southeast of France in the county of Var. These soils originated from siliceous and calcareous mineral parent materials, respectively. Their characteristics were previously described by Criquet et al. (2007), and some among the principal ones are given in table 1. Soil samplings were performed on the 28th of January 2003 between vine rows, from the 0 to 10 cm layer, which was devoid of vegetation at sampling time.

Table 1. Physical and chemical characteristics of soils and sewage sludge compost

Characteristic[a]	Soil 1 ("Pourcieux")	Soil 2 ("La Garde Freinet")	Sewage sludge compost
Clay (%)	22.6	12.9	
Silt (%)	40.3	20.0	
Sand (%)	36.0	67.0	
pH (H_2O)	8.55	6.9	7.0
Organic matter (g kg^{-1})	8.8	11.83	508
Organic C (g kg^{-1})	5.12	6.88	254
Total N (g kg^{-1})	0.6	2.31	21.8
Total CaCO3 (g kg^{-1})	319	0	

[a] Data refer to dry matter (105°C) of surface soil (0-10 cm) sieved at < 2 mm. For additional data see Criquet et al. (2007) and Criquet and Braud (2008).

2.2. Origin and Characteristics of the Sewage Sludge Compost (SSC)

Mature sewage sludge compost (SSC) used during the experiment was purchased on the 28th of January 2003 from Biotechna Company, located at Ensuès-la-Redonne (Bouches-du-Rhône, France). It was obtained by mixing green wastes (175 m^3) with biological sludge (70 m^3) for 180 d, and its maturity was certified conform to the French Standard (AFNOR, 2002). SSC characteristics were previously described by Criquet et al. (2008), and some among the principal ones are given in table 1. Furthermore, other chemical characteristics of this SSC, obtained by near-infrared reflectance spectroscopy (NIRS) and reflecting humification status of SSC, are given by Albrecht et al. (2008).

2.3. Incubation Experiments: Chemical and Biological Analyses

For the incubation experiments, the soils were previously sieved (< 2 mm) and thereafter mixed with SSC or with P-salts constituted by KH_2PO_4 or NaH_2PO_4 salts. Organic and mineral amendments were carried out in order to provide to soil a P total quantity of 0.22 g kg^{-1} by soil (equivalent to 0.5 g P_2O_5), a value which is coherent with those mentioned by Pote et al. (2003) and Ormeno et al. (2005) and which represents an exacerbated rate commonly used in experimental protocols. In the case of SSC, this corresponded to an application rate of 92 T ha^{-1}, a value that has ever been commonly used in other studies about the effect of SSC on soil properties (Ormeno et al., 2005; Selivanovskaya and Latypova, 2006). The different mixtures were distributed in 9x9x9 cm polypropylene pots and overlaid with perforated black plastic to avoid plant growth. Each pot (mesocosm) contained 750 g of an individually prepared mixture and the experiments were carried out in triplicate. Distilled water was added to each soil mixture to bring it to 60% of its water holding capacity, and water content of each mixture was adjusted every week. Soils without amendments were used as controls. Pots were incubated in the open air in the experimental field of the Faculty of Sciences of Saint-Jérôme (Marseille, France). Field incubation was performed over the period comprised between the 14th of February 2007 and the 12th of May of the same year. Climatic conditions over this period were representative of the French Mediterranean bioclimate (sub-humid). Mean air temperature of the region varied between 6°C (February) and 17°C (May), and monthly mean precipitations over the experimental period ranged between 5 and 80 mm (Météo France, for details see Larchevêque et al., 2005) Chemical and biological analyses were carried out after 25 and 87 days of incubation. Available P (P Olsen) was extracted according to Olsen et al. (1954), and PO_4^{3-} in the extracts were measured using the method of Murphy and Riley (1962). Results were expressed in mg of P Olsen g^{-1} dry weight (DW). Three types of phosphatase activities were measured in triplicate in each pot: (1) acid phosphomonoesterase (APH); (2) alkaline phosphomonoesterase (BPH); (3) phosphodiester-ase (PDI). Methods were based on the measurement of *p*-nitrophenol released during the incubation of soil with specific *p*-nitrophenyl phosphate esters (phosphate mono- or diester). Additionally, BPH/APH ratios were calculated using values of both BPH and APH activities (Dick et al., 2000). APH, BPH and PDI activities were measured according to the methods of Tabatabai and Bremner (1969) and Browman and Tabatabai (1978). Results were expressed in units defined as μmoles of p-nitrophenol released min^{-1} (U) g^{-1} dry weight (U g^{-1} DW). Culturable heterotrophic bacteria (CHB) were counted using a most-probable-number (MPN) method adapted from Joner et al. (2002). Results were expressed as log of bacteria g^{-1} dry weight (DW). Details of all the chemical and biological analyses are given by Criquet et al. (2007). All the analyses were carried out in triplicate for each pot.

2.4. Statistical Analyses

Means and standard error of means were determined for each set of triplicates. The means were compared using Kolmogorov-Smirnov test, with a significance level of $P < 0.05$. Since homogeneity of variances was not ensured, the non-parametric Kruskal-Wallis H tests were used to test significant effects of the different factors (kinds of fertilization, incubation time, soil and cation effects) on chemical and biological variables. To investigate soil effect on microbial responses and P availability, average control values were subtracted to data of the corresponding amended soils, in order to avoid interfering effect due to the fluctuations of the variables within controls over incubation time. Significant effects of the factors on the dependent variables (i.e., P Olsen, bacteria, APH, BPH, PDI and BPH/APH ratio) were retained for $P < 0.05$, $P < 0.01$ and $P < 0.001$. In order to reduce the dimensionality of data, a Principal-Components Analysis (PCA) was performed and treatments showing similar evolution in their characteristics were clustered using a Hierarchic Ascendant Classification (HAC). All calculations were performed using Statistica software version 6.0 (StatSoft, Maisons-Alfort, France).

3. Results

3.1. Available P

The range of variation of available P in the soils subjected to different amendment modalities was comprised between 13.9 µg and 112.2 µg P g^{-1} dry weight. The lowest available P concentrations were detected in controls without amendments over all the duration of the incubation experiment. All the amendments resulted in significant increases in available P in both calcareous and siliceous soils (table 2). After 25 d of incubation, soils amended with both Na and K phosphate salts showed higher available P contents with regards to SSC amended ones. Increasing incubation time to 87 d resulted in most of cases in a significant enhancement of available P in amended as well as in control soils (figure 1, table 2). Increases in available P in controls between the two dates were 17.3 µg and 14.5 µg g^{-1} DW in calcareous and siliceous soils respectively. Subtracting these values to those obtained at 87 d in fertilized soils allowed determining the origin of these increases in each kind of treatment. Following this subtraction, it appeared (data not detailed) that the available P concentrations were not significantly different between the two dates in both Na and K-phosphate salt fertilized pots. On the contrary, P contents in SSC pots still showed significant increases between 25d and 87 d. Indeed, SSC amended soils showed the higher increases in available P between the two dates. Moreover, these increases were more important in siliceous soil, and this was also true for the soils amended with phosphate salts. Thus, as also mentioned in the table 2, the mineral parent material of soils had a significant effect on the evolution of available P. On the contrary, the nature of the cation within the different P salts had no effect on the evolution of P Olsen in soils (table 2).

Table 2. Results of Kruskal-Wallis H test

	Variables					
Effect	APH	BPH	BPH/APH	PDI	Bact	P Olsen
Sewage sludge compost	0.65^{ns}	0.96^{ns}	2.43^{ns}	0.75^{ns}	0.003^{ns}	8.33**
Phosphate salt	2.78^{ns}	0.25^{ns}	5.42*	2.18^{ns}	0.55^{ns}	23.35***
Cation	11.26^{ns}	20.44^{ns}	12.44^{ns}	11.29^{ns}	16.05^{ns}	19.07^{ns}
Incubation time	14.84***	29.41***	3.75*	1.19^{ns}	0.14^{ns}	5.72*
Soil type	0.627^{ns}	5.48*	3.92*	2.12^{ns}	2.21^{ns}	4.36*

*, **, ***: significant at $P < 0.05$, $P < 0.01$ and $P < 0.001$ respectively. ns: not significant.

Data refer to H values of non-parametric Kruskal-Wallis test. APH: acid phosphatase; BPH: alkaline phosphatase; PDI: phosphodiesterase; Bact: culturable heterotrophic bacteria; P Olsen: available P concentration.

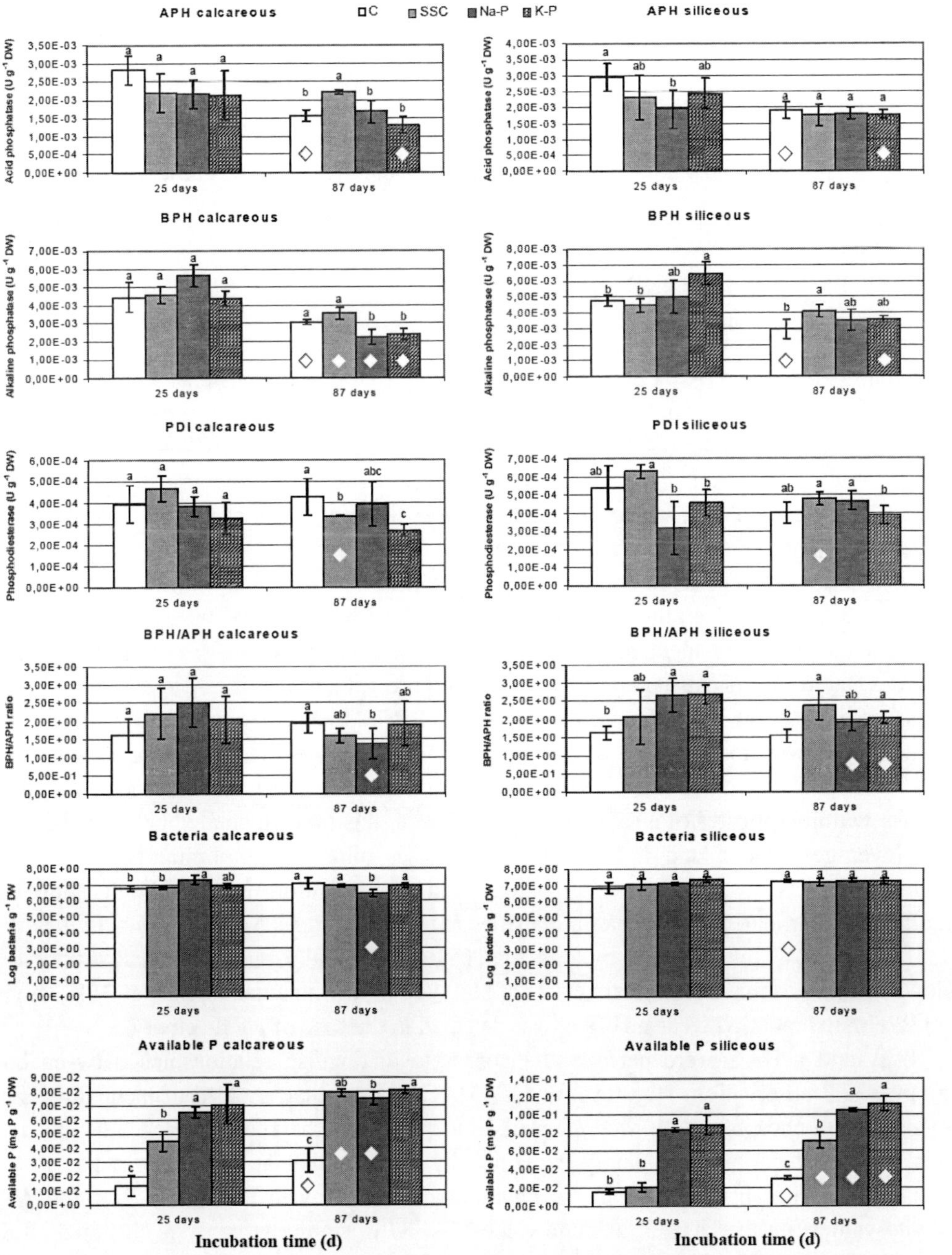

Figure 1. Acid (APH) and alkaline (BPH) phosphomonesterase and phosphodiesterase (PDI) activities, BPH/APH ratio, bacteria density and available P in amended and control soils after 25 and 87 days of incubation. (C) control; (SSC) Sewage sludge compost; (Na-P & K-P) Na and K phosphate salts. Bars indicate standard errors of means (n = 3). Values with the same letters are not significantly different ($P < 0.05$). (◇) Symbols indicate significant differences across both sampling dates for each treatment ($P < 0.05$).

3.2. Bacteria

The average numbers of culturable heterotrophic bacteria ranged between 6.44 and 7.31 log units (figure 1). Results showed a very low variability in their numbers during the whole experiment. As a consequence, SSC and phosphate salts, cation type, incubation time and soil type did not globally show any significant effect on bacteria densities as described in table 2.

3.3. Phosphatase Activities

Three kinds of phosphatase activities were measured, acid and alkaline phosphomonoesterases (APH and BPH respectively) and phosphodiesterases (PDI). The last one always showed lower activities with regards to the phosphomonoesterases. Indeed, PDI activities were about 4 to 16-fold inferior to both APH and BPH activities. As revealed by Kruskal-Wallis tests (table 2), soil amendments did not globally show any significant effect (*i.e.* SSC, P-salts and cation type of P-salts) on the different pools of phosphatase activities. However, a trend to decrease in PDI activities was observed at 25 d in phosphate salts amended soils with regards to controls and SSC soils. PDI activities following the different treatment were also generally not significantly affected by neither incubation time nor soil type. On the contrary, incubation time had a significant effect on both APH and BPH activities (table 2), since as it can be seen on the figure 1, their activities decreased during the course of incubation. As a consequence, some BPH/APH ratios were also significantly affected by the incubation time effect (figure 1, table 2). However, it must be noted that BPH/APH ratio showed important fluctuations with regards to treatment and incubation time, but also within replicates of a same treatment. Indeed, this ratio fluctuated between 1.38 and 2.69 (average values), and a trend in its increase following both mineral and organic amendments was observed, especially after 25 d of incubation. Moreover, this ratio was significantly correlated with bacteria densities ($r = 0.61$, $P < 0.05$) and BPH activities ($r = 0.71$, $P < 0.05$), as revealed also by PCA map (figure 2). Finally, statistical analyses (table 2) allowed distinguishing a significant effect of soil type following amendments on BPH activity and BPH/APH ratio, whereas no effect was detected in the case of APH activity.

PCA and a HAC were performed in order to distinguish relationships between both variables and observations (figures 3 and 4). HAC showed that, at 25 d of incubation, soils amended with phosphate salts were clustered independently than controls and SSC amended soils. This segregation was mainly explained by the fact that at 25 d, Na and K P-salts amended soils had high P available contents, and often showed the lowest PDI and APH activity values (on the basis of average values). At 87 d, similarity analysis clustered all the observations in a third and single group, distinct from the others, and thus confirmed the effect of incubation time on most of the variables.

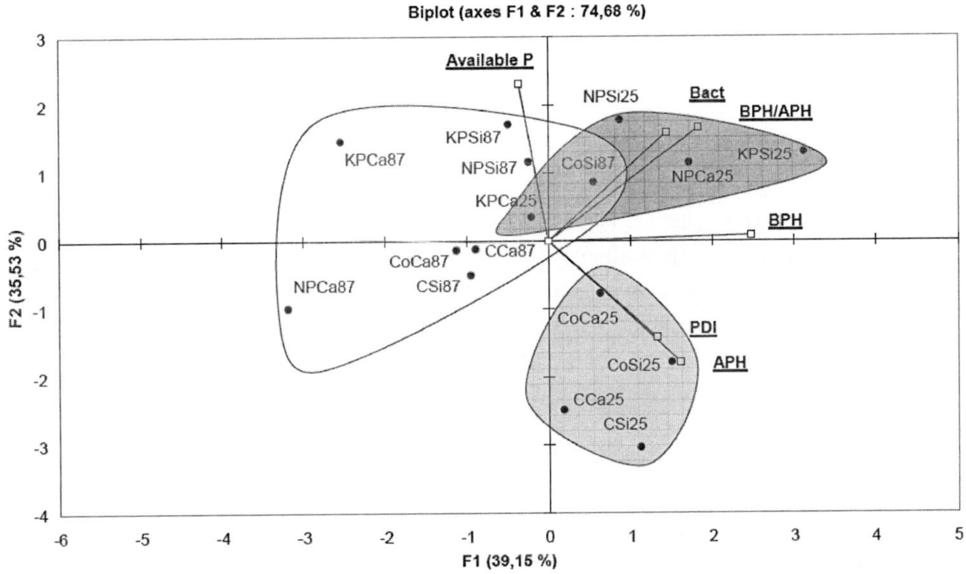

Figure 2. Principal component analysis of the variables studied in amended and control soils after 25 and 87 days of incubation. (APH) Acid phosphatase; (BPH) Alkaline phosphatase; (PDI) Phosphodiesterase; (Bact) Culturable heterotrophic bacteria. (Available P) Available P concentration. C, Co, NaP and KP refer to amendment modalities, i.e., control, sewage sludge compost, Na and K-phosphate salts respectively. 25 and 87 (days) refer to the incubation time. Ca and Si refer to parent mineral material of soils, i.e., calcareous and siliceous respectively.

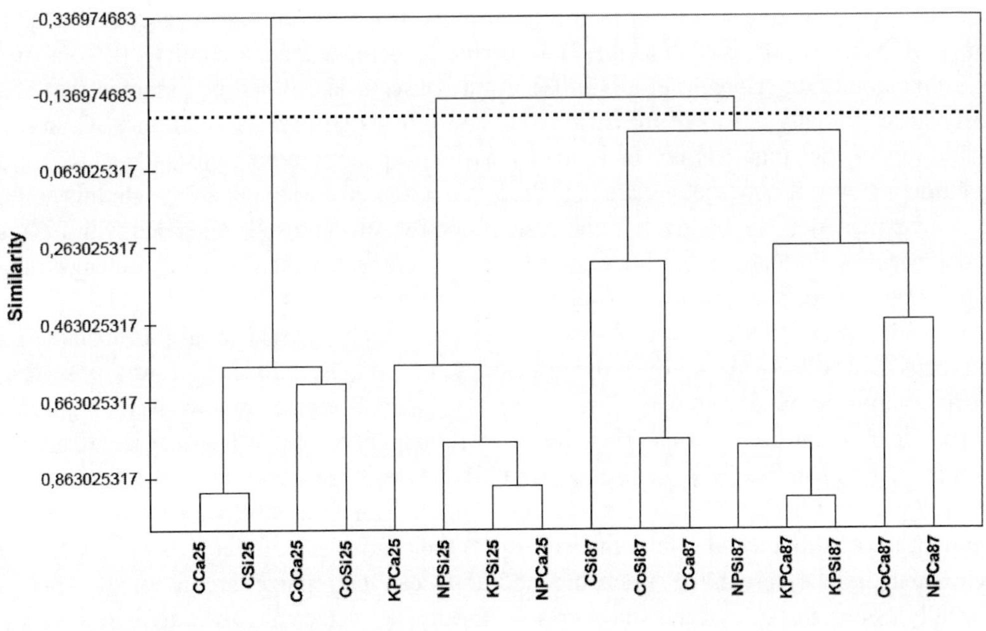

Figure 3. Hierarchic ascendant classification of the different treatments. Legends are given in figure 2.

4. Discussion

Fertilizing both calcareous and siliceous soils with sewage sludge compost (SSC) or mineral phosphate salts resulted in low or no significant modifications in biological variables with regard to controls without amendment. On the contrary, all of the amendments used as P sources resulted in significant increases in available P contents of soils. Moreover, increasing the incubation time had a significant effect on most of the variables investigated. These observations matched exactly with our previous study performed on a very degraded and nonagricultural soil subjected, among other things, to the same amendments. Other studies regarding the effects of municipal waste compost were carried out by some of our colleagues (Larchevêque et al., 2005, 2006; Ormeno et al., 2006) in order to investigate the effect of SSC on soil properties under natural field conditions. They confirmed our results obtained with mesocosms, since they showed also that SSC increased available P concentration as well as other soil fertility parameters (e.g., SOM, CEC, total N, etc.), and they did not observe any significant modifications in either bacteria or fungi densities. Thus, all of these convergent observations, each performed on different soils with different physico-chemical characteristics, allow to the finding of an improving SSC effect on soil P fertility, even if this improvement still remains difficult to explain because of its lack of relationships with microbial parameters, i.e., microorganisms and P mineralizing enzyme activities. In previous studies (Criquet et al., 2007; Criquet and Braud, 2008), we focused on the effects of sewage sludge (SS), without composting with green wastes, on P availability and microbial parameters involved in organic matter P mineralization. In this case, we found that SS considerably enhanced soil phosphatase activities, whereas this was not the case with SSC. We explained these differences by the effect of C substrates within these kinds of amendments. In SS, they are easily available and thus they can promote microbial growth and high microbial P demand resulting in high phosphatase excretion by microorganisms. On the other hand, the availability of C for microorganisms is low in mature SSC because of its high degree of humification, which makes it difficult to degrade and assimilate (Sikora and Yakovchenko, 1996; Albrecht et al., 2008). As a consequence, it can be assumed that soil phosphatases initially adsorbed in soil or SSC constituents still release P from SSC or soil organic matter, but this P is not or is poorly assimilated by microorganisms because of C starvation and thus it may accumulate in soils, as we observed during our study. On this point, the same explanation can be given in the case of control soils and soils amended with P-salts in conjunction with temperature increase over the experiment, which probably enhanced the hydrolyzing rates of phosphatases between the two incubation dates.

Considering soil effect, we observed that the mineral parent origin of soils (i.e., calcareous and siliceous) significantly affected some variables. Indeed, it was observed, especially in soils amended with P-salts, that available P concentrations were higher in siliceous than in calcareous soils. This can be easily explained by the fact that in calcareous soil it is well known that PO_4^{3-} can form insoluble salts with Ca^{2+} ions, and this can decrease the available P fraction (Santruckova et al., 2004). This last observation may also be important for environmental safety. Indeed, part of the available P can represent a risk for the environment, since it can be leachable and thus can eutrophize groundwater. This is especially true in the case of the siliceous soil, because we detected high water extractable P concentrations which were about 2–3 times higher than in the calcareous one (data not shown). Soil type also affected BPH activity and therefore BPH/APH ratio. Even if this effect was less evident than with available P, BPH and BPH/APH ratio tended to be higher in the siliceous soil. According to Dick et al. (2000) and Nèble et al. (2007), alkaline soils show generally higher BPH activity. This was not true in our study, since the pH of the siliceous soil (6.9) was lower than that of the calcareous soil (8.55). Thus, another explanation of this soil effect could be the chemical characteristics of the soils, since siliceous soil had higher initial nutrient concentrations than the calcareous one, which may improve microbial metabolism. Also, soil may exhibit micro-sites within which pH can be higher with regard to bulk soil, and this may enhance excretion and activity of some enzymes.

Results showed also that bacteria density was significantly correlated with BPH/APH ratio, and previous studies have mentioned that plants are not known to produce BPH (Tarafdar and Claassen, 1988; Guillemin et al., 1995). Thus, this may indicate that bacteria density was more effective than other soil organisms in producing BPH and, therefore, in modifying BPH/APH ratio, as also mentioned by Criquet et al. (2004) and Alarcon-Gutierrez (2007).

A last enzyme activity, that of phosphodiesterase, was also investigated during this study, and its activity followed similar pattern with regards to the other ones during the incubation experiments. However, this phosphatase had considerably lower activity than phosphomono-esterases (i.e., BPH and APH), and it can be hypothesized that this probably reflected the fact that BPH and APH substrates (phosphomonoesters) were more abundant than those of PDI (phosphodiesters) in soils, as also mentioned by Turner and Haygarth (2005). Moreover, considering the fact that phosphodiesterases in soils can also originate from plants, Asmar and Gissel-Nielsen (1997) also pointed out that phosphomonoesterases excreted by plant roots were always predominant with regard to phosphodiesterases.

5. Conclusion

Results obtained in the present study confirmed those obtained in previous ones (Larchevêque et al., 2005; 2006; Ormeno et al., 2006; Criquet et al., 2007; Criquet and Braud, 2008) performed using quite different soils, and under different experimental conditions (i.e., mesocosm and field studies). They showed that sewage sludge compost as well as P-salt fertilization did not globally affect microbial parameters involved in P recycling from organic matter. On the contrary, available P was even increased in all treatments, including controls, and we suggested that C availability probably interacted with the phenomena observed in P nutrient mineralization, as also argued by Allison and Vitousek (2005). These convergent results are of particular importance, since these observations indicate that phenomena following these kinds of P fertilization may be generalized to a large range of soils showing different physico-chemical characteristics. Moreover, it also suggests that phosphatase activities, as well as microbial densities, are not suitable bioindicators of the short-term effects of these agricultural practices on soil properties.

References

AFNOR. (2002). French Standard U 44-095. Organic soils improvers-composts containing substances essential to agriculture, stemming from water treatment, AFNOR Edition, Saint-Denis-la-Plaine, France.

Alarcon-Gutierrez, E. (2007). Influence de facteurs abiotiques sur la régulation des paramètres microbiens impliqués dans la dégradation de la matière organique d'une litière forestière méditerranéenne. PhD Thesis, University Paul Cézanne, Marseille, France, 226 pp.

Albrecht, R., Joffre, R., Gros, R., Le Petit, J., Terrom, G., & Périssol, C. (2008). Efficiency of near-infrared reflectance spectroscopy to assess and predict the stage of transformation of organic matter in the composting process. *Bioresource Technol.* 99, 448-455.

Allison, S.D., & Vitousek, P.M. (2005). Responses of extracellular enzymes to simple and complex nutrient inputs. *Soil Biol. Biochem.* 37, 937-944.

Asmar, F., & Gissel-Nielsen, G. (1997). Extracellular phosphomono- and phosphodiesterase associated with and released by the roots of barley genotypes: a non-destructive method for the measurement of the extracellular enzymes of roots. Biol. Fertil. Soils 25, 281-289.

Browman, M.G., & Tabatabai, M.A. (1978). Phosphodiesterase activity of soils. *Soil Sci. Soc. Am. J.* 42, 284-290.

Caravaca, F., Masciandaro, G., & Ceccanti, B. (2002). Land use in relation to soil chemical and biochemical properties in a semiarid Mediterranean environment. *Soil Till. Res.* 68, 23-30.

Criquet, S., Ferre, E., Farnet, A.M., & Le Petit, J. (2004). Annual dynamics of phosphatase activities in an evergreen oak litter: influence of biotic and abiotic factors. *Soil Biol. Biochem.* 36, 1111-1118.

Criquet, S., Braud, A., & Nèble, S. (2007). Short-term effects of sewage sludge application on phosphatase activities and available P fractions in Mediterranean soils. *Soil Biol. Biochem.* 39, 921-929.

Criquet, S., & Braud, A. (2008). Effects of organic and mineral amendments on available P and phosphatase activities in a degraded Mediterranean soil under short-term incubation experiment. *Soil Till. Res.* 98, 164-174.

Dick, W.A., Cheng, L., & Wang, P. (2000). Soil acid and alkaline phosphatase activity as pH adjustment indicators. *Soil Biol. Biochem.* 32, 1915-1919.

European Commission, Council Directive 86/278/EEC of 12 June 1986 on the protection of the environment, and in particular of the soil, when sewage sludge is used in agriculture. *Official Journal L.* 181 (1986) pp 6-12.

Guillemin, J.P., Orozco, M.O., Gianinazzi-Pearson, V., & Gianinazzi, S. (1995). Influence of phosphate fertilization on fungal alkaline phosphatase and succinate dehydrogenase activities in arbuscular mycorrhiza of soybean and pineapple. *Agr. Eco. Environ.* 53, 63-69.

Hooda, P.S., Truesdale, V.W., Edwards, A.C., Withers, P.J.A., Aitken, M.N., Miller, A., & Rendell, A.R. (2001). Manuring and fertilization effects on phosphorus accumulation in soils and potential environmental implications. *Adv. Environ. Res.* 5, 13-21.

Joner, E.J., Corgié, S.C., Amellal, N., & Leyval, C. (2002). Nutritional constraints to degradation of polycyclic aromatic hydrocarbons in a simulated rhizosphere. *Soil Biol. Biochem.* 34, 859-864.

Larchevêque, M., Baldy V., Korboulewsky, N., Ormeno, E., & Fernandez, C. (2005). Compost effect on bacterial and fungal colonization of kermes oak leaf litter in a terrestrial Mediterranean ecosystem. *Appl. Soil Ecol.* 30, 79-89.

Larchevêque, M., Ballini, C., Korboulewsky, N., & Montès, N. (2006). The use of compost in afforestation of Mediterranean areas : effects on soil properties and young tree seedlings. *Sci. Total Environ.* 369, 220-230.

Lima, J.A., Nahas, E., & Gomes, A.C. (1996). Microbial populations and activities in sewage sludge and phosphate fertilizer-amended soil. *Appl. Soil Ecol.* 4, 75-82.

Melero, S., Porras, J.C.R, Herencia, J.F., & Madejon, E. (2006). Chemical and biochemical properties in a silty loam soil under conventional and organic management. *Soil Till. Res.* 90, 162-170.

Murphy, J., & Riley, J.R. (1962). A modified single solution method for the determination of phosphate in natural waters. *Anal. Chem.* 27, 31-36.

Nèble, S., Calvert, V., Le Petit, J., & Criquet, S. (2007). Dynamic of phosphatase activities in a cork oak litter (Quercus suber L.) following sewage sludge application. *Soil Biol. Biochem.* 39, 2735-2742.

Olsen, S.R., Cole, C.V., Watanabe, F.S., & Dean, L.A. (1954). Estimation of available phosphorus in soils by extraction with sodium bicarbonate. United States Department of Agriculture, Circular 939. United States Government Printing Office, Washington DC, USA.

Ormeno, E., Baldy, V., Ballini., C, Larchevêque, M., Périssol, C., & Fernandez, C. (2006). Effects of environmental factors and leaf chemistry on leaf litter colonization by fungi in a Mediterranean shrubland. *Pedobiologia.* 50, 1-10.

Pote, D.H., Lory, J.A., & Zhang, H. (2003). Does initial soil P level affect water-extractable soil P response to applied P? *Adv. Environ. Res.* 7, 503-509.

Saggar, S., Parfitt, R.L., Salt, G., & Skinner, M.F. (1998). Carbon and phosphorus transformations during decomposition of pine forest floor with different phosphorus status. *Biol. Fertil. Soils.* 27, 197-204.

Santruckova, H., Vrba, J., Picek, T., & Kopacek, J. (2004). Soil biochemical activity and phosphorus transformations and losses from acidified forest soils. *Soil . Biol. Biochem.* 36, 1569-1576.

Selivanovskaya, S.Y., & Latypova, V.Z. (2006). Effects of composted sewage sludge on microbial biomass, activity and pine seedling in nursery forest. *Waste Management.* 26, 1253-1258.

Sikora, L.J., & Yakovchenko, V. (1996). Soil organic matter mineralization after compost amendment. *Soil Sci. Soc. Am. J.* 60, 1401-1404.

Speir, T.W., Horswell, J., Van Schaik, A.P., McLaren, R.G., & Fietje, G. (2004). Composted biosolids enhance fertility of a sandy loam soil under dairy pasture. *Biol. Fertil. Soils.* 40, 349-358.

Tabatabai, M.A. (1994). Soil enzymes, In: Methods of Soil Analysis, Part 2, Microbiological and Biochemical Properties, Soil Science Society of America, Madison, pp 775-833.

Tabatabai, M.A., & Bremner, J.M. (1969). Use of *p*-nitrophenyl phosphate for assay of soil phosphatase activity. *Soil Biol. Biochem.* 1, 301-307.

Tarafdar, J.C., & Claassen, N. (1988). Organic phosphorus compounds as a phosphorus source for higher plants through the activity of phosphatases produced by plant roots and microorganisms. *Biol. Fertil. Soils*. 5, 308-312.

Turner, B.L., & Haygarth, P.M. (2005). Phosphatase activity in temperate pasture soils: Potential regulation of labile organic phosphorus turnover by phosphodiesterase activity. *Sci. Total Environ.* 344, 27-36.

In: Soil Fertility
Editors: D. P. Lucero and J. E. Boggs

ISBN: 978-1-60741-466-05

Chapter 5

CONTRIBUTION OF THE ROOT CAP TO SOIL FERTILITY: EXTRACELLULAR PLANT LECTINS

G. Curlango-Rivera[1], G. Albala[1], J.P. Kemp[2], D.V. Duclos[1] and M.C. Hawes[1]

[1] Department of Plant Sciences, Division of Plant Pathology and Microbiology, University of Arizona, Tucson AZ 85721

[2] Department of Biology, Luther College, Decorah IA 52101

ABSTRACT

Lectins are proteins with multiple sites that bind specific sugars and therefore can agglutinate cells or other substrates with those sugars on the surface. This property is exploited in detecting human ABO blood type, defined based on the presence of distinct surface sugars that result in agglutination of blood cells in response to appropriate lectins. This capacity to bind to specific sites on cell surfaces is of interest for its potential in treating human diseases including cancer, AIDS, and diabetes. In plants, species-specific lectins are secreted from root tips into the extracellular environment of agronomically important crops including legumes and cereals. The function, stability and distribution of these proteins after export from root cells into the extracellular environment remain unexplored. Experiments to examine predictions of the hypothesis that root-secreted lectins influence soil structure and fertility, rhizosphere microbial communities, and root nutrient cycling are warranted.

ACKNOWLEDGMENTS

This study was funded in part by a fellowship to GC-R, from CONSEJO NACIONAL DE CIENCIA Y TECNOLOGIA (CONACYT). A preliminary report of this work was presented at the Undergraduate Biology Research Conference, January 2007, University of Arizona (*Role of extracellular lectins in eucaryotic defense: Root border cells as a model;* authors: Hillebrand E, Kemp J, Olsen D, Curlango-Rivera G, Wen F, Pierson LS, Hawes MC).

Introduction: Influence of the Root Cap on Soil Properties

The root cap is the leading edge of the root system [Aiken and Smucker 1996; Barlow 2003; Feldman 1984]. As newly generated root tissues penetrate the soil, the arrival of the root cap at a given point at a given time results in fundamental changes in soil fertility in the immediate vicinity of the root tip [Curl and Truelove 1986; Foster et al 1983]. The root cap is programmed to create dramatic changes in the chemical and physical properties of the emergent environment by the delivery of a complex mixture of polysaccharides, proteins, soluble exudates [Bacic et al 1986; Chaboud 1983; Knee et al 2001; Moody et al 1988; Rougier et al 1979; Wright 1975]. Also released are detached living cells called root border cells [Hawes et al. 1998; Odell et al. 2008; Somasundaram et al 2008]. Overall, fifty percent or more of total fixed carbon is released from roots into the soil environment [Farrar et al 2001; Whipps and Lynch 1990] Direct measurements of the dry weight of material collected by washing axenic roots into water have shown that, in young healthy seedlings without wounds, >90% of the total carbon released from roots into the soil environment is derived from the root tip [Gunawardena et al 2002].

The dynamic nature of this process of releasing materials from the root cap into the surrounding environment can be observed by watching a root tip as it is placed into water, over the course of 1-2 minutes using a dissecting microscope (Figure 1). When removed from environments with high humidity but no free water, as are present in most soils most of the time, the root tip surface appears to be smooth (Figure 1A, inset). Within 20-30 seconds after immersing into water, a mass of material can be seen to swell away from the tip as the mucilage takes up water to 1000 times its weight (Figure 1A) [Sealey et al 1995]. Moving the tip gently to one side yields a stream of material in its wake (Figure 1B, arrows). Taking a photo at the moment a tap is administered to the plate results in swirling release throughout the water droplet (Figure 1C). When the root is removed from the water, thousands of individual cells are left behind in suspension (Figure 1D).

Root Border Cells

These root 'border' cells (Figure 1D) are uniquely differentiated in comparison with other plant cells, by virtue of their natural morphology as a population of detached cells delivered into the soil environment [Hawes et al 2000]. In addition to their unusual morphology, border cells also exhibit unique patterns of gene expression: As soon as they detach from the root cap, border cells exhibit protein and mRNA profiles that are distinct from other root tissues and from progenitor cells within the root cap [Brigham et al 1995; 1998]. Cell viability can be measured using vital stains like neutral red or fluorescein diacetate, or by observation of cytoplasmic streaming or metabolism of glucose [Hawes and Wheeler 1982; Stubbs et al 2004]. How long the cells survive depends on temperature, species, genotype, culture conditions and other environmental and genetic factors [Ijima et al 2000, 2004]. In some species, like sunflower and *Arabidopsis thaliana*, most cells from the root cap die before they detach into the surrounding environment [Hawes and Pueppke 1986]. In crop species like cereals, legumes, cotton, and cucurbits, by contrast, border cells are >95% viable when they are released from the root cap, and exhibit a high rate of metabolic activity [Brigham et al 1995; Stubbs et al 2004]. In hydroponic conditions, border cells from pea and corn can remain 100% viable over a period of months [Knudson 1919]. In roots of maize plants grown in field soil, border cells can survive for a week or more after detachment from the cap [Vermeer and McCully 1982].

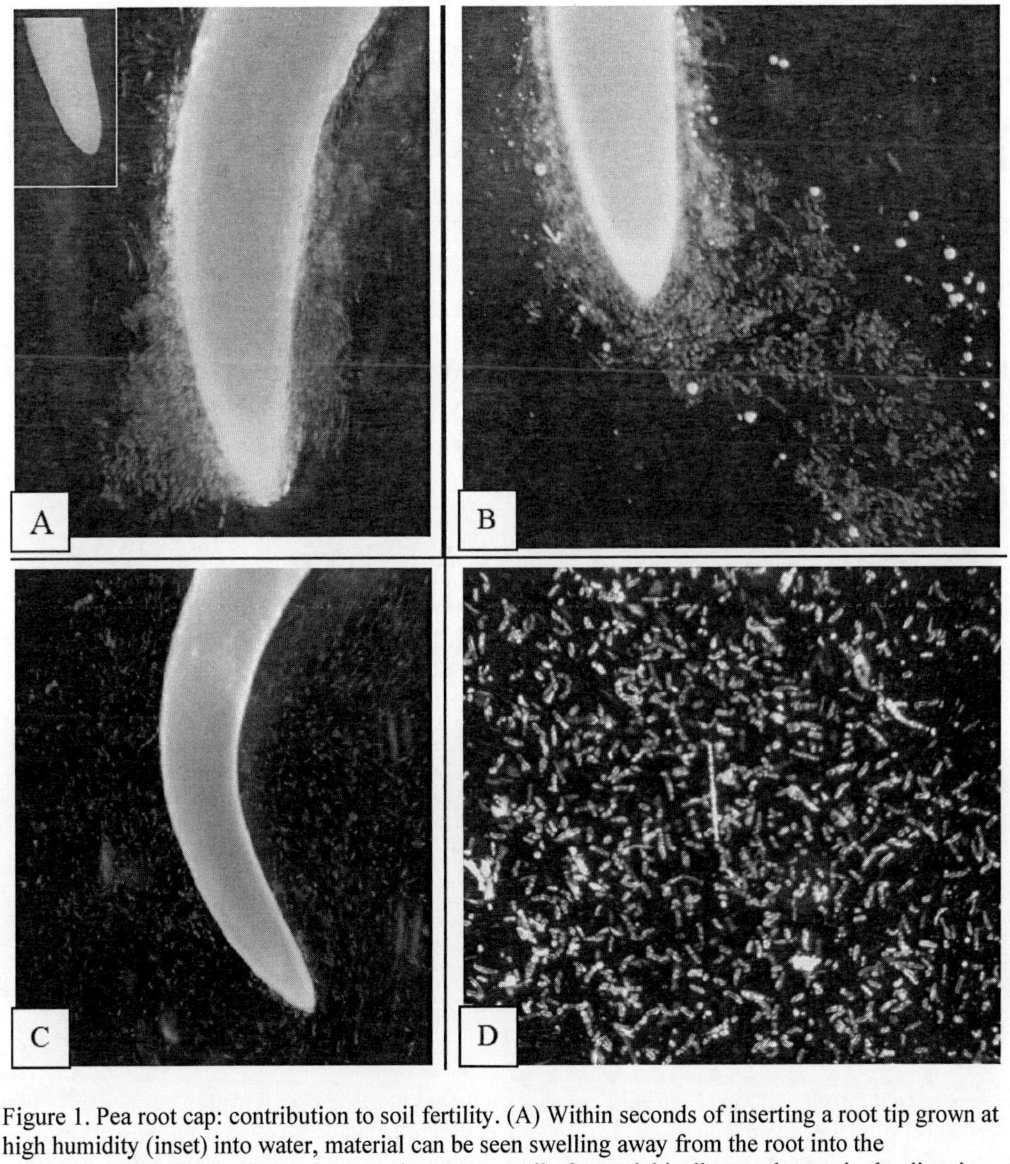

Figure 1. Pea root cap: contribution to soil fertility. (A) Within seconds of inserting a root tip grown at high humidity (inset) into water, material can be seen swelling away from the root into the environment; (B) When dragged across the stage, a trail of material is dispersed away in the direction of movement; (C) With gentle agitation, all material disperses away from the root surface into water; (D) At higher magnification, the extruded material can be seen to be a dispersed population consisting of hundreds of living, detached cells [Hawes and Wheeler 1982].

The viability and metabolic activity of the populations of detached cells is important, because they actively secrete a range of diverse nutrient-rich materials that can be predicted to influence soil fertility in a highly localized and specific manner. For example, pea border cells maintain an extracellular pH <5.0, a phenomenon which by itself could influence a range of soil properties including solubilization of phosphate and uptake of added fertilizers [Wen et al 1999; Gunawardena et al 2002]. Border cells also can specifically agglutinate soilborne bacteria into tightly localized populations [Hawes and Pueppke 1987, 1989]. Within 60 minutes of co-cultivation in water, in the absence of added nutrients, virtually all border cells and bacteria are removed from suspension as they co-aggregate into tight clumps that cannot be dispersed readily (Figure 2A). Certain bacteria, but not others, stimulate the production of contiguous strands that link cells and microorganisms even at a distance (Figure 2B). The nature of the interaction is such that the aggregated state is maintained over the course of days, in the absence of added nutrients [Hawes and Pueppke 1987, 1989].

Fungal and bacterial pathogens, symbionts, and saprophytes can be stimulated to germinate and grow in response to signals from border cells [Fusconi 1983; Gochnauer et al 1990; Gunawardena et al 2002; Nagahashi and Douds 2004]. In some cases, however, the detached cells release substances that can inhibit growth over a period of days [Gunawardena et al 2005]. The chemical nature of the inhibitory factor(s), or the basis by which it can remain active for such an extended period, is not known.

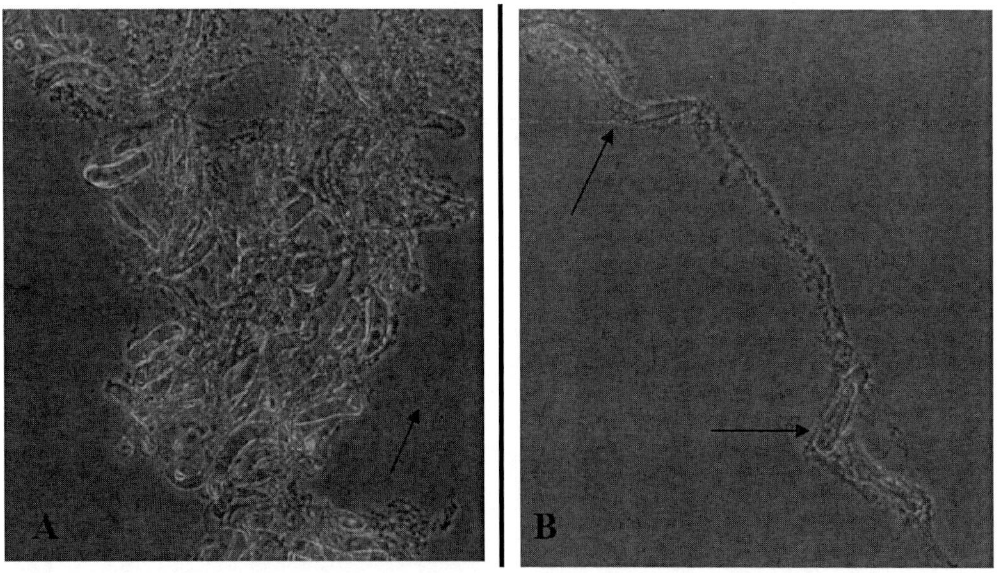

Figure 2. Agglutination of *Agrobacterium tumefaciens* by pea border cells. (A) When pea border cells @ 5000 cells/ ml are mixed with bacteria @ 10^5 /ml, both types of cells are removed from suspension within 1-2 h of incubation, and are aggregated into tight bundles that precipitate, leaving the supernatant with only a few free bacteria evident (arrow); (B) Individual border cells (arrows) are linked by strands of bacteria [Hawes and Pueppke 1987, 1989].

LECTINS IN THE SOIL ENVIRONMENT

This capacity of border cells to aggregate soilborne microorganisms is likely to play a significant role in defining the properties of the soil environment, but the mechanism of how the cell-cell interaction occurs is unclear. The material surrounding border cells contains a complex polysaccharide matrix comprised of fucose, glucose, galactose, and other sugars in diverse linkage groups, together with secreted proteins [Bacic et al 1986; Chaboud 1983; Knee et al 2001; Matsuyama et al 2001; Park et al 2002; Moody et al 1988; Rougier et al 1979; Wright 1975]. In pea, our primary model system, more than one hundred proteins are synthesized and secreted as border cells separate and disperse into the environment [Brigham et al 1995]. Recent studies have shown that the proteins, as a group, play a role in the aggregation properties of border cells and microbial populations [Wen et al 2007]. The specific proteins that condition this response are not yet clear, but one obvious candidate protein within the pea root cap secretome is pea lectin (PSL) (Figure 3). This protein has biochemical properties that would be predicted to facilitate agglutination and aggregation phenotypes, and to exhibit the stability needed to maintain its activity after secretion from living cells.

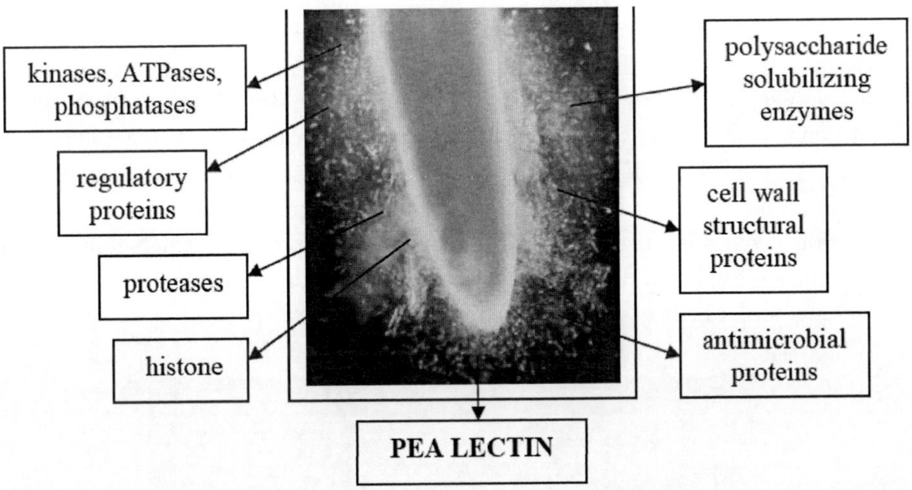

Figure 3. Classes of pea root cap secretome proteins (from Wen et al, 2007).

Lectins were identified >100 years ago and their defining property--the capacity to bind to specific sugars--and its many potential applications, were recognized early [Bouckaert et al 1999; Chrispeels and Raikhel 1991; Shakirova and Bezrukova 2007]. This property involves the capacity to alter the function of molecules with a particular matching configuration at the surface resulting, for example, in cellular agglutination, precipitation of glycoproteins, and altered surface chemistry [Peumans and VanDamme 1995; Rudiger and Gabius 2001]. The diverse ways lectins can be expected to be useful in human health, based on these biochemical properties, are illustrated in current articles describing efforts to identify new sources of lectins and to use them to control diseases ranging from HIV, cancer, and diabetes [e.g., Chandra et al 2006; Andre et al 2007; Balzarini et al 2006, 2007; Favacho et al 2007; Gilchrist et al 2006; Heinrich et al 2005; Moore et al 2007]. Efforts to use lectins in food safety assays and to develop disease and insect resistant crops also are ongoing [Coelho et al 2007; Cox et al 2006; Down et al 2006; Edwards and Singh 2006; Giovanini et al 2007; Hossain et al 2006; Kanrar et al 2002; Kim et al 2006; Lehrman et al 2007; Macedo et al 2007; Saha et al 2006; Singh et al 2006; Soares et al 2007].

The actual function(s) of any given lectin in its source plant remain unclear, but a study in which expression of endogenous legume lectins was suppressed has highlighted how important lectins may prove to be in plant health [Brill et al 2001]. Expression of two alfalfa lectin encoding genes, *MsLEC1* and *MsLEC2*, was reduced using antisense mRNA to each gene. The results revealed that lectins play unexpectedly important roles in growth regulation: abnormalities in tissue proliferation and organ morphology occurred in embryos, seedlings and plantlets, pod set was dramatically reduced, and seed lethality was increased by threefold.

LECTINS: A ROLE IN AGGREGATION AT THE ROOT-SOIL INTERFACE?

In roots, bulk expression of lectin is low, but localization studies have shown that its presence is concentrated in specific sites including the tips of root hairs and the extracellular mucilage surrounding root tips [Buffard et al 1988; Diaz et al 1984]. Lectins localized in root hair tips can influence the development of Rhizobium-legume symbiosis by enhancing microbial colonization [Diaz et al 1995; reviewed in Hirsch 1999]. Extracellular lectin has been identified in root exudates of wheat, clover, bean, and pea [Buffard et al 1988; Diaz et al 1984; Kjemtrup et al 1995; Mishkind et al 1983a,b]. In wheat, immunolocalization was used to show that the lectin from wheat--called wheat germ agglutinin (WGA)--is secreted from emergent root caps within embryos (Figure 4). This protein, which binds N-acetyl glucosamine (GlcNAc), continues to be secreted as the radicle emerges, and ultimately is secreted among border cells in older roots [Mishkind et al 1983b].

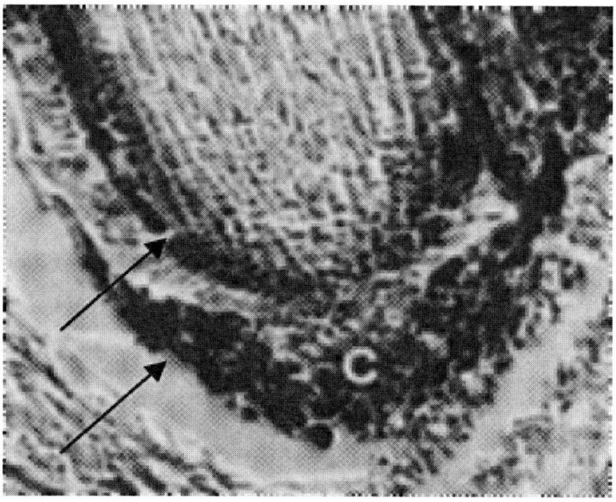

Figure 4. Wheat germ agglutinin localized in peripheral cells of the root cap in wheat embryo. Immunolocalization reveals wheat germ agglutinin (black stain, arrows) in emerging root cap (c) of wheat embryo (from Mishkind et al 1983, used with permission).

The functional impact of extracellular WGA as it is released into the environment of emerging root tips is not known. One possible effect, assuming that GlcNAc binding sites are present on the surfaces of extracellular components, is to hold the mixture of border cells and matrix polysaccharides into a aggregated mixture. The available surface sugars on border cell walls are not known, but the monosaccharide compositions of their surrounding mucilages have been profiled and found to contain variable levels of rhamnose, glucose, galactose, mannose, xylose, fucose, arabinose and amino sugars [Bacic et al 1986; Chaboud 1983; Knee et al 2001; Moody et al 1988; Rougier et al 1979; Wright 1975]. If border cells of wheat carry a binding site for WGA, then the presence of this protein could in part explain why border cell aggregation occurs under some circumstances. We tested this by adapting the standard assay used to evaluate human ABO blood type for application with populations of detached border cells (Figure 5). The results revealed that WGA added to border cells does indeed result in aggregation of the cells (Figure 5). Addition of the hapten sugar, GlcNAc, resulted in immediate dispersal of the agglutinated cells, but other sugars had no effect. A survey of border cells of other cereals, legumes, and cucurbits in combination with their corresponding lectins revealed similar results. The capacity to hold border cells and their associated mucilages together as a network, as they detach from the root that produces them, at a minimum could be predicted to influence fertility by controlling the distribution of nutrients within the soil environment.

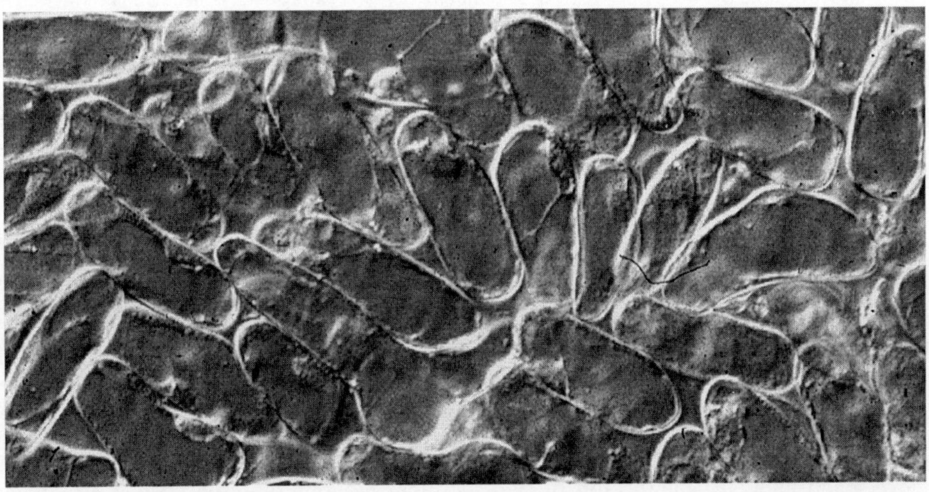

Figure 5. A plant cell 'haemagglutination' assay. Wheat border cells (~700 cells in 50 microliters) were added to wells of a flat bottom microtitre plate. WGA (5 microliters @ 5 mg/ml) was added. Within 5 minutes, cells aggregated into a ball that became more cohesive over a 15 minute period into a network of cells that bounced as a unit when agitated. Addition of a few grains of GlcNAc, the hapten sugar, resulted in instantaneous dispersal of the cells (as in Figure 1D). Other sugars including mannose, glucose, xylose, galactose, and cellobiose had no effect on the agglutination when added to the WGA-border cell network.

Lectins in Defense of the Growing Root Tip

Speculation that lectins play a role in plant defense has been based largely on observations that lectins are ubiquitous, have great variability in binding specificity, exhibit localized expression in vegetative tissues, and can be toxic to eucaryotes and procaryotes when applied in laboratory assays [Chrispeels and Raikhel 1992; Peumans and Van Damme 1995; Rudiger and Gabius 2001; Shakirova and Bezrukova 2007]. Soilborne pathogens have been shown to be specifically agglutinated by specific lectins, which in some cases results in growth inhibition [Barkai-Golan et al. 1975; Christinzio et al. 1988]. Potentially the most important impact lectins could have in soil fertility, apart from fostering symbiosis between plants and nitrogen-fixing bacteria, is in protection of growing root tips and their delivery of nutrient-rich materials as they move through soil [Mishkind et al. 1983a,b]. To date, the hypothesis that lectins function in defense remains untested. However, ectopic expression of various lectins has provided promising results for crop improvement, along the same lines as expression of other selective poisons like *Bacillus thuringiensis* (BT) toxin. Efforts to exploit lectins in the control of insect pests is an especially strong focus [Coelho et al 2007; Down et al 2006; Edwards and Singh 2006; Giovanini et al 2007; Hossain et al 2006; Kim et al 2006; Macedo et al 2007; Saha et al 2006; Singh et al 2006; Soares et al 2007]. Transgenic plants with increased resistance to fungal pathogens have been reported by expressing lectins toxic to fungi--orchid lectin and rubber tree lectin (hevein), respectively--in tobacco and Indian mustard [Cox et al 2006; Kanrar et al 2002]. Some authors have speculated that lectins will most likely be effective against pests that invade host cells and ingest the poison as cells are disrupted and cytoplasmic and vacuolar contents are dispersed [Chrispeels and Raikhel 2001].

Mishkind et al [1983] were the first to propose that root cap secreted lectins might function in root defense against pathogen attack. We used pea lectin to examine predictions of this hypothesis. Pea lectin is among the best-characterized plant lectins, thanks to its historical role as a model system in the Rhizobium-legume lectin hypothesis [Diaz et al. 1984, 1995; Hirsch 1999]. Most proteins within the pea root cap secretome (Figure 3), including polysaccharide-solubilizing and other enzymes as well as cell wall and defense proteins, are members of large multigene families. Pea lectin, in contrast, is encoded by a single gene (psl) in the *P. sativum* genome [Hoedemaeker et al 1994]. This can facilitate genetic approaches, such as gene silencing, that will allow dissection of its role in defense of roots. The protein is an ellipsoidal, nonglycosylated dimeric lectin with two sugar-binding sites, for glucose and mannose, each located near the top of two identical monomers. How such a protein might influence the soil environment as roots penetrate into new territory is unknown. Based on an extensive literature search and interviews with colleagues, the possibility that pea lectin can inhibit microbial growth does not appear to have been examined previously. There is preliminary evidence to suggest that it can inhibit growth of pollen beetle larvae, when expressed transgenically in oilseed rape [Lehrman et al 2007].

When pea lectin was added to a population of spores from the pea pathogen, *Nectria haematococca*, the spores were agglutinated into a tight mass (Figure 6A). No germination occurred over the course of 24-48 h. In contrast, in control populations without pea lectin spores were dispersed throughout the suspension and 100% germination occurred within 24 h (Figure 6B). These data indicate that glucose and/or mannose binding sites are present on the surface of these fungal spores, allowing agglutination to occur in response to pea lectin. There are two possible explanations for the inhibition of spore germination: The lectin could be toxic to the spores, or the agglutination per se could result in inhibition. *N. haematococca* spores produce a self-inhibitor, such that germination is prevented when local spore concentrations are high. Either mechanism could have a significant impact on the ability of spores to germinate and infect their hosts, in the natural soil environment.

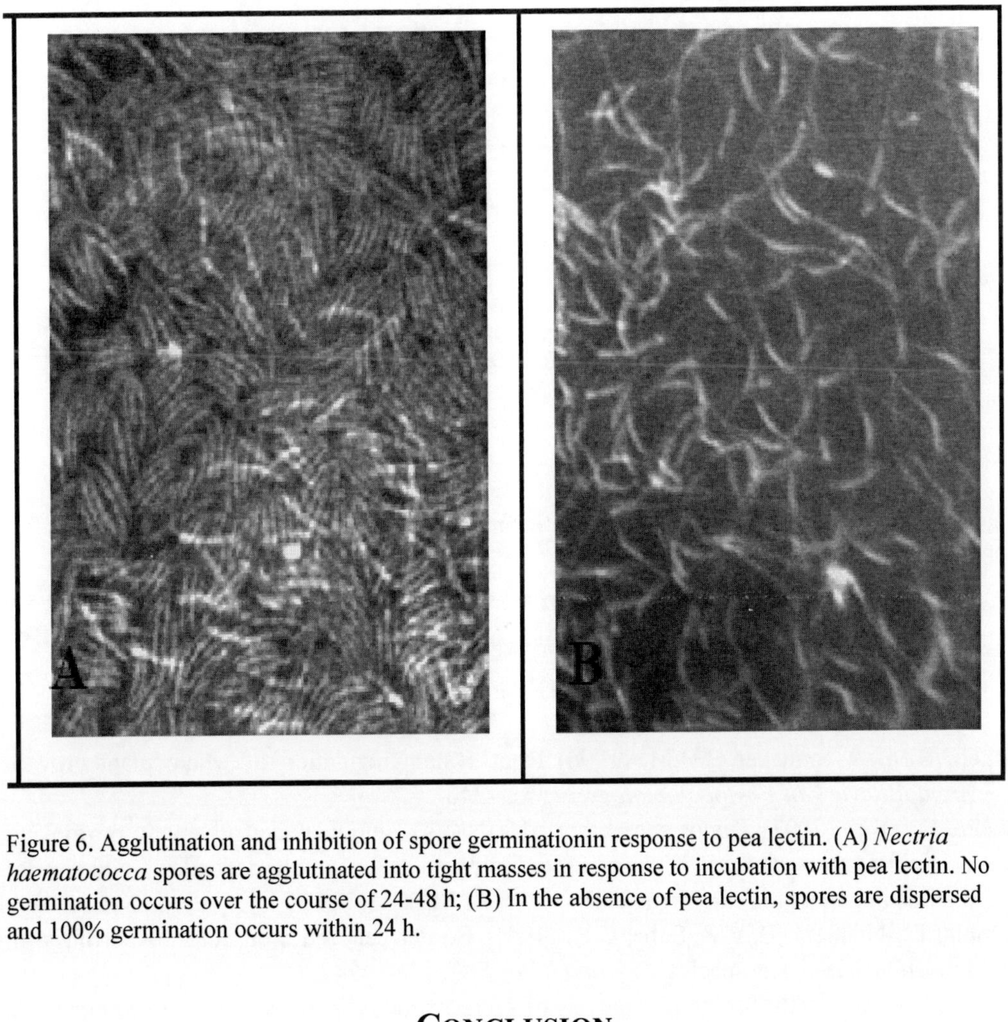

Figure 6. Agglutination and inhibition of spore germinationin response to pea lectin. (A) *Nectria haematococca* spores are agglutinated into tight masses in response to incubation with pea lectin. No germination occurs over the course of 24-48 h; (B) In the absence of pea lectin, spores are dispersed and 100% germination occurs within 24 h.

CONCLUSION

A review of lectin biology offers hints of ways that lectin delivery might influence soil fertility, and ways these proteins might be exploited in crop production (Figure 7). Binding of lectins to rhizosphere microflora, soil, soil and plant additives potentially could have positive or negative effects on plant productivity, depending on genetic and environmental conditions [Barkai-Golan et al. 1978; Christinzio et al. 1988; Graber et al. 2006; Hendrie 1976; Hirsch 1999; Wu and Liu 2008]. These biological activities might be especially adaptable to manipulation in controlled environment agriculture where the root system can be monitored and accessed directly.

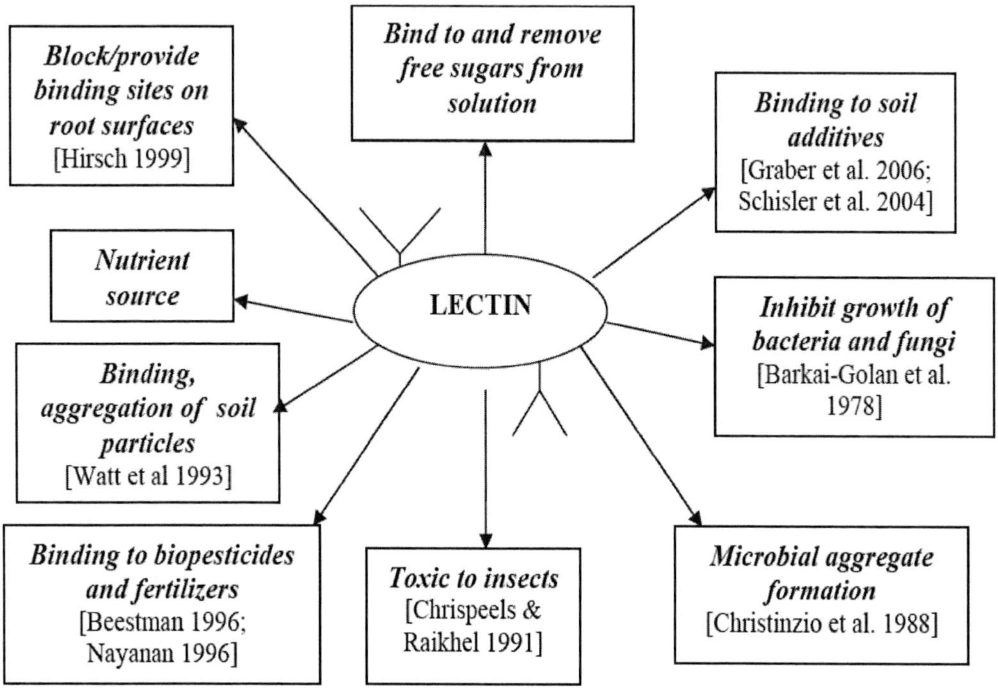

Figure 7. Extracellular lectins delivered into the environment by the root cap: Possible functional impact in soil fertility.

References

Aiken, R.M. & Smucker, A.J.M. (1996) Root system regulation of whole plant growth. *Annual Review of Phytopathololology 34:* 325-346.

Andre, S., et al. (2007) Tumor suppressor p16(INK4a) - modulator of glycomic profile and galectin-1 expression to increase susceptibility to carbohydrate-dependent induction of anoikis in pancreatic carcinoma cells. *FEBS J 274*: 3233-3256.

Arriola, L., Niemera, B.A. & Safir, G.R. (1997) Border cell and arbuscular mycorrhizae in four *Amaranthaceae* species. *Phytopathology 87*: 1240-1242.

Bacic, A., et al. (1986) Structural analysis of secreted root slime from maize (*Zea mays* L.). *Plant Physiology 80*: 771-777.

Balzarini J, et al. (2007) Carbohydrate-binding agents efficiently prevent dendritic cell-specific intercellular adhesion molecule-3-grabbing nonintegrin (DC-SIGN)-directed HIV-1 transmission to T lymphocytes. *Molecular Pharmacology 71:* 3-11

Balzarini, J. (2006) Inhibition of HIV entry by carbohydrate-binding proteins. *Antiviral Research 71*: 237-247.

Barkai-Golan, R., et al. (1978) Studies on growth inhibition by lectins of Penicillia and Aspergilli. *Archives of Microbiology 116:* 119-124.

Barlow, P.W. (2003) The root cap: Cell dynamics, cell differentiation and cap function. *Journal of Plant Growth Regulation 21*: 261-286.

Bouckaert, J., et al. (1999) Novel structures of plant lectins and their complexes with carbohydrates. *Current Opinion in Structural Biology 9*: 572-577.

Beestman, G.B. (1996) Emerging technology: the bases for new generations of pesticide formulation. In: Pesticide formulation and adjuvant technology. Edited by C.L. Foy and D. W. Pritchard. CRC Press USA. pp 43-68.

Brigham, L.A., et al. (1995) Differential expression of proteins and mRNAs from border cells and root tips of pea. *Plant Physiology 109:* 457-463.

Brill, L.M. et al. (2001) Expression of MsLEC1- and MsLEC2-antisense genes in alfalfa plant lines causes severe embryogenic, developmental and reproductive abnormalities. *Plant Journal 25:* 453-461.

Buffard, D., et al (1988) Lectin-gene expression in pea roots. *Planta 173*: 367-372.

Chaboud, A. (1983) Isolation, purification and chemical composition of maize root cap slime. *Plant and Soil 73:* 395-402.

Chandra, N.R., et al. (2006) Lectindb: a plant lectin database. *Glycobiology 16:* 938-946.

Chrispeels, M.J. & Raikhel, N.V. (1991) Lectins, lectin genes, and their role in plant defense. *Plant Cell 3*: 1-9.

Cristinzio, G., et al. (1988) Agglutination response of the conidia of eight *Fusarium* species to lectins having different sugar-binding specifities. *Plant Pathology 37*: 120-124.

Coelho, M.B., et al. (2007) Insecticidal action of *Annona coriacea* lectin against the flour moth *Anagasta kuehniella* and the rice moth *Corcyra cephalonica* (Lepidoptera : Pyralidae). *Bulletin of Entomological Research 96*: 43-52.

Cox, K.D., et al. (2006) Gastrodia anti-fungal protein from the orchid *Gastrodia elata* confers disease resistance to root pathogens in transgenic tobacco. *Planta 224*: 1373-1383.

Curl, E.A, & Truelove, B. (1986) The Rhizosphere. Berlin: Springer-Verlag. 288 pp.

Diaz, C.L., et al. (1984) Determination of pea root lectin using ELISA. *Planta 161*: 302-307.

Diaz, C.L. et al, (1995) Sugar-binding activity of pea lectin expressed in white clover hairy roots. *Plant Physiology 109:* 1167-1177.

Down RE, et al. (2006) Insecticidal spider venom toxin fused to snowdrop lectin is toxic to the peach-potato aphid, *Myzus persicae* (Hemiptera : Aphididae) and the rice brown planthopper, *Nilaparvata lugens* (Hemiptera : Delphacidae). *Pest Management Science 62:* 77-85.

Edwards, O. & Singh, K.B. (2006) Resistance to insect pests: What do legumes have to offer? *Euphytica 147*: 273-285.

Farrar, J., et al. (2003) How roots control the flux of carbon to the rhizosphere. *Ecology 84:* 827–837.

Favacho, A.R.M., et al. (2007) In vitro activity evaluation of *Parkia pendula* seed lectin against human cytomegalovirus and herpes virus 6. *Biologicals 35*: 189-194

Feldman, L.J. (1984) Development and dynamics of the root apical meristem. *American Journal of Botany 7:* 1308-1314.

Foster, R.C., et al. (1983) Ultrastructure of the root-soil interface. St. Paul, MN: APS Press.

Fusconi, A. (1983) The development of the fungal sheath on *Cistus incanus* short roots. *Canadian Journal of Botany 61:* 2546-2553.

Gilchrist, C.A., et al. (2006) Impact of intestinal colonization and invasion on the *Entamoeba histolytica* transcriptome. *Molecular and Biochemical Parasitology 147*: 163-176.

Giovanini, M.P., et al. (2007) A novel wheat gene encoding a putative chitin-binding lectin is associated with resistance against Hessian fly. *Molecular Plant Pathology 8:* 69-82.

Gochnauer, M.B., et al. (1990) Do detached root-cap cells influence bacteria associated with maize roots? *Plant Cell and Environment 13*: 793-801.

Gouget, A., et al. (2006) Lectin receptor kinases participate in protein-protein interactions to mediate plasma membrane-cell wall adhesions in Arabidopsis. *Plant Physiology 140:* 81-90.

Graber, E.R., et al. (2006) Soil stabilization in semiarid and arid land agriculture. *Journal of Materials in Civil Engineering 18*: 190-205.

Guinel, F.C. & McCully, M.E. (1987) The cells shed by the root cap of zea: their origin and some structural and physiological properties. *Plant, Cell and Enviornment 10:* 565-578.

Gunawardena, U. & Hawes, M.C. (2002) Role of border cells in localized root infection by pathogenic fungi. *Molecular Plant Microbe Interactions 15:* 1128-1136.

Gunawardena, U, et al. (2005) Tissue specific localization of pea (*Pisum sativum* L.) root infection by *Nectria haematococca*: Mechanisms and consequences. *Plant Physiology 137:* 1363-1374.

Hawes, M.C., et al. (1998) Function of root border cells in plant health: Pioneers in the rhizosphere. *Annual Review of Phytopathology 36:* 311-327.

Hawes, M.C., et al. (2000) The role of root border cells in plant defense. *Trends in Plant Science 5*: 128-133.

Hawes, M.C., et al. (2003) Root caps and rhizosphere. *Journal of Plant Growth Regulation* 21: 352-367.

Hawes, M.C., Pueppke SG (1986) Sloughed peripheral root cap cells: yield from different species and callus formation from single cells. *American Journal of Botany 73*: 1466-1473.

Hawes, M.C., Pueppke, S.G. (1987) Correlation between binding of *Agrobacterium tumefaciens* by root cap cells and susceptibility of plants to crown gall. *Plant Cell Reports 6*: 287-290.

Hawes, M.C., Pueppke, S.G. (1989) Reduced rhizosphere colonization ability of *Agrobacterium tumefaciens* chromosomal virulence mutants. *Plant and Soil 113:* 129-132.

Hawes, M.C., Wheeler H.E. (1982) Factors affecting victorin-induced cell death. *Physiological Plant Pathology 20:* 137-144.

Heinrich, E.L., et al. (2005) Direct targeting of cancer cells: A multiparameter approach. *Acta Histochemistry 107*: 335-344.

Hendrie, R.A. (1976) Granulated fertilizers. Noyes Data Corporation USA 339 p.

Hirsch, A.M. (1999) Role of lectins (and rhizobial exopolysaccharides) in legume nodulation. *Current Opinion in Plant Biology 2:* 320-326.

Hoedemaeker, F.J. et al. (1994) Pea seed isolectins 1 and 2 and pea root lectin result from carboxypeptidase-like processing of a single gene product. *Plant Molecular Biology 24:* 75-81.

Hossain, M.A., et al. (2006) Transgenic expression of onion leaf lectin gene in Indian mustard offers protection against aphid colonization. *Crop Science 46:* 2022-2032.

Ijiima, M.B. et al. (2000) Sloughing of cap cells and carbon exudation from maize seedling roots in compacted sand. *New Phytologist 145*: 477-482.

Ijima, M.P., et al (2004) Method to quantify root border cells in sandy soil. *Soil Biology and Biochemistry 36:* 1517-1519.

Jiang, J. et al. (2007) Overexpression of OsJAC1, a lectin gene, suppresses coleoptile and stem elongation in rice. *Journal of International Plant Biology 49*: 230-237

Kanrar, S., et al. (2002) Transgenic expression of hevein, the rubber tree lectin, in Indian mustard confers protection against *Alternaria brassicae*. *Plant Science 162*: 441-448.

Kehr, J. (2006) Phloem sap proteins: their identities and potential roles in the interaction between plants and phloem-feeding insects. *Journal of Experimental Botany 57*: 767-774.

Kilpatrick, D.C. (1999) Mechanisms and assessment of lectin-mediated mitogenesis. *Molecular Biotechnology 11*: 55-65.

Kim, C.G., et al. (2006) Monitoring the occurrence of genetically modified soybean and maize around cultivated fields and at a grain receiving port in Korea. *Journal of Plant Biology 49:* 218-223.

Kjemtrup, S., et al. (1995) Targeting and release of phytohemagglutinin from the roots of bean seedlings. *Plant Physiology 109:* 603-610.

Knee, E.M., et al. (2001) Root mucilage from pea and its utilization by rhizosphere bacteria as a sole carbon source. *Molecular Plant-Microbe Interactions 14:* 775-784.

Knox, J.P. (1992) Molecular probes for the plant cell surface. *Protoplasma 167:* 1-9.

Knudson, L. (1919) Viability of detached root cap cells. *American Journal of Botany 6*: 309-410.

Komath, S.S., et al. (2006) Beyond carbohydrate binding: new directions in plant lectin research. *Organic Biomolecular Chemistry 4:* 973-988.

Kozlov, Y.V., et al. (2006) Ribosome-inactivating lectins of plants. *Molecular Biology 40*: 711-723.

Lannoo, N., et al. (2006) The presence of jasmonate-inducible lectin genes in some but not all *Nicotiana* species explains a marked intragenus difference in plant responses to hormone treatment. *Journal of Experimental Botany 57*: 3145-3155.

Lehrman, A, et al. (2007) Influence of pea lectin expressed transgenically in oilseed rape (*Brassica napus*) on adult pollen beetle (*Meligethes aeneus*). *Journal of Applied Entomology 131:* 319-325.

Lynch, M.M. & Whipps, J.M. (1990) Substrate flow in the rhizosphere. *Plant and Soil 129*: 1-10.

Macedo, M.L.R., et al. (2007) Insecticidal action of *Bauhinia monandra* leaf lectin (BmoLL) against *Anagasta kuehniella* (Lepidoptera : Pyralidae), *Zabrotes subfasciatus* and *Callosobruchus maculatus* (Coleoptera : Bruchidae). *Comparative Biochemistry and Physiology; Molecular and Integrative Physiology146*: 486-498.

Matsuyama, T et al. (1999) A maize glycine-rich protein in synthesized in the lateral root cap and accumulates in the mucilage. *Plant Physiology 120*: 665-674.

Mishkind, M.L. et al (1983) Localization of wheat germ agglutinin-like lectins in various species of the Gramineae. *Science 220*: 1290-1292.

Mishkind, M.L. et al (1983) The cell biology of wheat germ agglutinin and related lectins. Chemical Taxonomy, Molecular Biology and Function of Plant Lectins, p 163-176.

Moody SF, et al. (1988) Structural analysis of secreted slime from wheat and cowpea roots. *Phytochemistry 27:* 2857-2861.

Moore, J.S., et al. (2007) Reactivities of N-acetylgalactosamine-specific lectins with human IgA1 proteins. *Molecular Immunology 44*: 2598-2604.

Nagahashi, G. & Douds, D.D. (2004) Isolated root caps, border cells, and mucilage from host roots stimulate hyphal branching of the arbuscular mycorrhizal fungus, *Gigaspora gigantea. Mycological Research 108:* 1079-1088.

Nayayanan, K.S. (1996) Macro- and microemulsion technology and trends. In: Pesticide formulation and adjuvant technology. Edited by C.L. Foy CL and D.W. Pritchard. CRC Press USA. pp 115-174.

Odell, R.E. et al. (2008) Stage-dependent border cell and carbon flow from roots to rhizosphere. *American Journal of Botany 95:* 441-446.

Park, S.W., et al. (2002) Isolation and characterization of a novel ribosome-inactivating protein from root cultures of pokeweed and its mechanism of secretion from roots. *Plant Physiology 130*: 164-178.

Peumans, W.J., VanDamme, J.M. (1995) The role of lectins in plant defence. *Histochemical Journal 27:* 253-271.

Rougier, M., et al (1979) Use of lectin to detect the sugar components of maize root cap slime. *Journal of Histochemical Cytochemistry 27:* 878-881.

Rudiger, H, Gabius, HJ (2001) Plant lectins: Occurrence, biochemistry, functions and applications. *Glycoconjugate Journal 18*: 589-613.

Saha, P., et al. (2006) A novel approach for developing resistance in rice against phloem limited viruses by antagonizing the phloem feeding hemipteran vectors. *Plant Molecular Biology 62*: 735-752.

Sanmartin, M, et al. Divergent functions of VTI12 and VTI11 in trafficking to storage and lytic vacuoles in Arabidopsis. *PNAS 104*: 3645-3650.

Schisler, D.A., et al. (2004) Formulation of Bacillus spp. for biological control of plant diseases. *Phytopathology 94:* 1267-1271.

Sealey, L.J., McCully, M.E., Canny, M.J. (1995) The expansion of maize root cap mucilage during hydration. *Physiologia Plantarum 93*: 38-46.

Shakirova, FM, Bezrukova, MV. (2007) Current knowledge about presumable functions of plant lectins. *Zhurnal Obshchei Biologii 68:* 109-125.

Singh, K, et al. (2006) Exploration for anti-insect properties of lectin from seeds of soybean (*Glycine max*) using *Bactrocera cucurbitae* as a model. *Phytoparasitica 34*: 463-473.

Soares, EL, et al. (2007) Characterization and insecticidal properties of globulins and albumins from *Luetzelburgia auriculata* (Allemao) Ducke seeds towards *Callosobruchus maculatus* (F.) (Coleoptera : Bruchidae). *Journal of Stored Products Research 43:* 459-467.

Somasundaram, S., et al. (2008) Dynamics of root border cells in rhizosphere soil of *Zea mays* L.: crushed cells during root penetration, survival in soil, and long term soil compaction effect. *Plant Production Science 11*: 440-446.

Stubbs, V.E.C., et al. (2004) Root border cells take up and release glucose-C. *Annals of Botany 93:* 221-224.

Uren, N.C. (1993) Mucilage secretion and its interaction with soil, and contact reduction. *Plant and Soil 155*: 79-82.

van den Burg, H.A., et al. (2006) *Cladosporium fulvum* Avr4 protects fungal cell walls against hydrolysis by plant chitinases accumulating during infection. *Molecular Plant-Microbe Interacteractions 19*: 1420-1430.

Vermeer, J. & McCully, M.E.(1982) The rhizosphere in Zea: new insight into its structure and development. *Planta 156*: 45-51.

Watt, M., et al. (1993) Plant and bacterial mucilages of the maize rhizosphere: comparison of their soil binding properties and histochemistry in a model system. *Plant and Soil 151:* 151-165.

Wen, F. et al. (1999) Expression of an inducible pectinmethylesterase gene is required for border cell separation from roots of pea. *Plant Cell 11*: 1129-1140.

Wen, F, et al. (2007a) Extracellular proteins in *Pisum sativum* L. root tip and border cell exudates. *Plant Physiology 143*: 773-783.

Wen, F. et al. (2007b) Proteins among the polysaccharides: a new perspective on root cap *slime*. *Plant Signaling & Behavior 2*: 410-412.

Wright, K. (1975) Polysaccharides of root-cap slime from five maize varieties. *Phytochemistry 14*: 759-763.

Wu, L. & Liu, M. (2008) Preparation and characterization of cellulose acetate-coated compound fertilizer with controlled-release and water-retention. *Polymers for Advanced Technology 19*: 785-792.

In: Soil Fertility
Editors: D. P. Lucero and J. E. Boggs
ISBN: 978-1-60741-466-81

Chapter 6

CONTRIBUTIONS OF LEAF-CUTTING ANTS TO SOIL FERTILITY: CAUSES AND CONSEQUENCES

Alejandro G. Farji-Brener* and Mariana Tadey [1]
Laboratorio Ecotono-INIBIOMA-CRUB-UNComa-CONICET
Pasaje Gutiérrez 1125, (8400) Bariloche, Argentina.

ABSTRACT

Leaf-cutting ants modify soil fertility through two mechanisms. First, the building, enlargement, and maintenance of nests ants affect soil structure, porosity and density. Second, leafcutters collect and concentrate vegetal material inside their nests to maintain their fungus culture, the food for most of the colony. As a result of this process, ants generate a huge quantity of organic waste that is deposited in nest cavities or dumps on the soil surface. This organic waste is very rich in nutrients. The content of organic Carbon, Nitrogen, Potassium, Phosphorous, and Magnesium is, in average, between 20 and 50 times greater in refuse dumps than in adjacent, non-nest soils. The process of waste deposition, thus, notably increases the nutrient content of soils around nests. Consequently, plants inhabiting nest areas often show greater abundance, growth rate, foliar and root biomass, and reproduction rate than plants outside nest areas. This positive effect on plants might scale-up and affect the structure of vegetation assemblages and the balance between trees and herbs at landscape scale. Several factors affect the contribution of leaf-cutting ants to soil fertility. Particularly, extreme weather, low abundance of palatable vegetation for leafcutters, competition with introduced mammalian herbivores and frequent fires reduce ant foraging rates, nest abundance, and therefore their production of organic waste. Since leaf-cutting ant activity and nest density is strongly dependent on the availability of pioneer or ruderal plant species, the strength of their contribution to soil fertility could be more important in early successional environments and disturbed habitats. Ant-nest areas should be especially protected because they are hot spot of plant diversity and core of plant succession. From a restoration point of view, the nutrient-rich refuse dumps can be employed as natural, free-access and ecologically sustainable fertilizer to improve soil fertility in degraded areas.

* E-mail: alefarji@crub.uncoma.edu.ar

INTRODUCTION

Insects positively affect soil fertility in different ways (Schowalter 1981). The deposition of sugar honeydew by aphids beneath their host plant provides an extra energy source for free-living nitrogen-fixing bacteria (Owen 1980, Stadler *et al.* 1998). The deposition of frass from canopy insects increase soil total C, N, and the soil NH_4^+ pool (Lovett & Ruesnik 1995, Fonte & Schowalter 2005, Frost & Hunter 2006). Ants and termites improve soil nutrients and organic carbon through the accumulation of organic matter, and by the building of nest-mounds and foraging galleries (Salick *et al.* 1983, Bruyn & Conacher 1995, Dangerfield *et al.* 1998). However, leaf-cutting ants of the genus *Atta* and *Acromyrmex* are considered the most important soil-modifier insect group in Neotropical ecosystems (Cherrett 1989, Farji-Brener & Illes 2000). In this chapter, we discuss the causes and consequences of this process (figure 1). Specifically, we will (a) describe the mechanisms by which leafcutters modify soil fertility, (b) review the physical and chemical soil changes generated by the ants quantifying their magnitude, (c) illustrate how these changes affect plant demography, plant assemblages and landscape structure, and (d) discuss which factors affect the contribution of leaf-cutting ants to soil fertility. There are two ways through which leaf-cutting ants may modify soil fertility: the building and maintenance of their nests and the production and deposition of their organic waste products. The first process changes physical properties of surrounding soils, and the second increases their nutrients concentration.

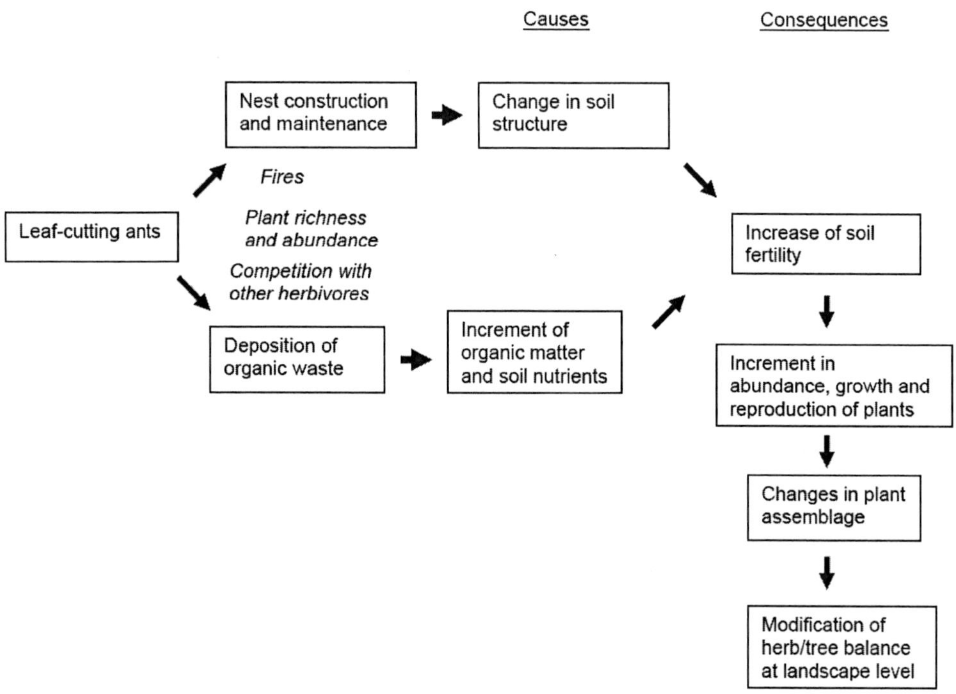

Figure 1. Schematic representation of how leaf-cutting ants improve soil fertility, the consequences of this soil nutrient improvement on vegetation, and the main factors affecting this process.

LEAF-CUTTING ANTS AFFECT SOIL STRUCTURE

The changes in physical soils properties caused by leaf-cutting ants are consequences of nest construction and maintenance. Leaf-cutting ant nests are formidable structures that may house millions of ants and their fungus colonies for 10-20 years (Weber 1982, Cherrett 1989). During the construction, expansion, and lifetimes of nests, ants construct deep underground chambers and move large quantities of soil to the nest surface. For example, an adult *Atta* nest can contain between 300 and 7800 chambers from near the surface to 7-m depth (Stahel & Geijskes 1939; Jonkman 1978; Moreira *et al.* 1995; Forti & Moreira 1997). To construct this underground network, ants may move up 23m^3 of soil (Autori 1942). All these ant activities alter the physical properties of soils (Cosarinsky & Roces 2007). In Costa Rican rain forests, *Atta cephalotes* modified 85% of the 27 soil profiles sampled in a 2.5 ha area. Soil materials were transported from the horizons AB and B to soil surface, making a new A1 soil horizon (Alvarado *et al.* 1981). This alteration of soil horizons through digging behavior and soil removal affects soil porosity, drainage and ventilation (Cherrett 1989). For example, soil resistance to penetration was significantly lower in nest-soils of *Atta sexdens* than in non-nest soils of the adjacent secondary forest. Nest-soils also showed lowest bulk density, higher macro-porosity and lower micro-porosity than non-nest soils (Mountinho *et al.* 2003). All these changes in soil structure affect soil biota activity, improving soil fertility around the nest area (Wagner *et al.* 1997, Boulton & Amberman 2006). Consequently, nest-soils are often more nutrient-rich than adjacent, control non-nest soils (Etter & Botero 1990, Farji-Brener & Silva 1995a, figure 2b).

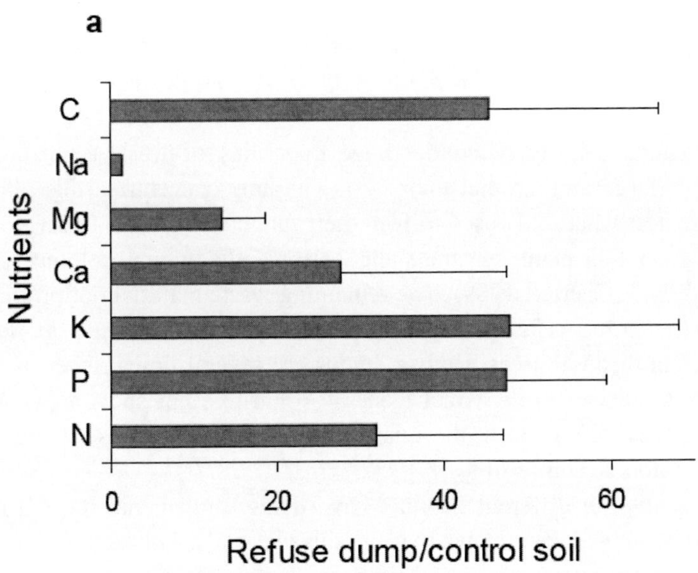

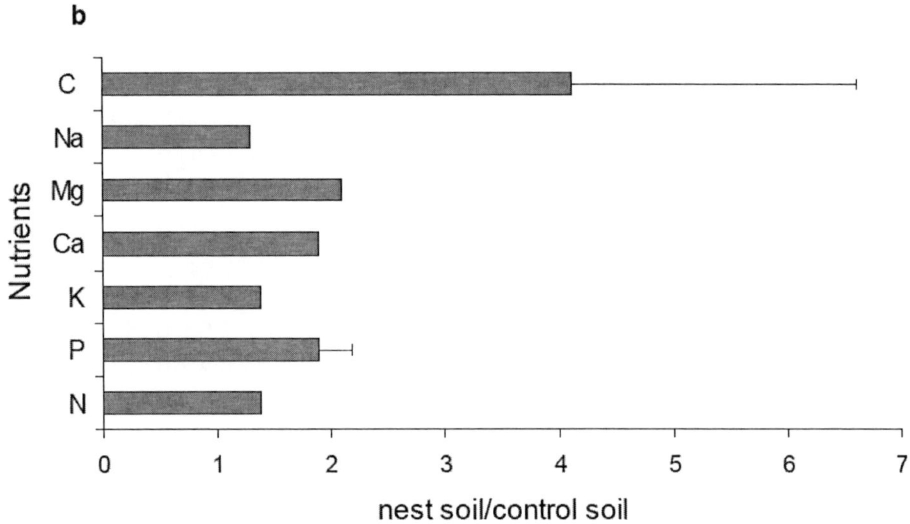

Figure 2. Contribution of leaf-cutting ants to soil fertility by refuse dumps (a) and ant-nest soils (b). X-axis represents the rate between the nutrient content of refuse dumps and control soils (a), and between nest-soils and control soils (b). Note that figures differ in the x-axis scale. Data were obtained from studies of *Atta leavigata* (nest soils) (Etter & Botero 1990, Farji-Brener & Silva 1995), *Atta sexdens* (refuse dumps) (Moutinho *et al.* 2003), *Atta Colombica* (refuse dumps) (Haines 1975, Wirth *et al.* 2003), and *Acromyrmex lobicornis* (refuse dumps) (Tadey & Farji-Brener 2007, Farji-Brener & Ghermandi 2008, Farji-Brener AG unpublished results). Values represent means ± 1 SE. C = Carbon, Na = Sodium, Mg = Magnesium, Ca = Calcium, K = Potassium, P = Phosphorous, and N = Nitrogen.

LEAF-CUTTING ANTS AFFECT SOIL NUTRIENT CONTENT

Leaf-cutting ants selectively collect huge quantities of fresh vegetation from plants growing within a large area around their nest. The ants concentrate this plant material by carrying it to the nest where it is used to feed their fungus gardens in underground chambers. The fungi grow on this plant substrate and parts of the fungal colonies are fed by the developing ant brood (Cherrett 1989). The remaining waste material not profitable for fungal decomposition (hereafter, refuse dumps) is removed from the fungus gardens to specific external or internal disposal areas. Refuse dumps are several times richer in organic carbon and nutrients than adjacent soils, which generates nutritive hot spots areas around the nest (Farji-Brener & Illes 2000). Both, the quality and quantity of refuse dumps generated are important contributors to soil fertility.

Numerous studies in different habitats have demonstrated that the nutrient content in refuse dumps is several times greater than in adjacent soils. For example, refuse dumps of leaf-cutting ant nests may have up to 80 times more nutrient content than adjacent non-nest soils in habitats ranging from tropical rain forests (Haines 1975, 1978; Moutinho, *et al.* 2003, Wirth, *et al.* 2003) to temperate deserts (Farji-Brener & Ghermandi 2000, 2004, 2008; Tadey & Farji-Brener 2007). However, not all soil nutrients were equally improved by the ants; a revision of the scientific literature reveals that the magnitude of the soil enrichment depends on the leaf-cutting ant species and the habitat studied (Table 1). While Na content was similar between refuse dumps and control soils, refuse dumps showed a mean higher content of Mg

(15-fold), Ca and N (25-fold) and C, P and N (50-fold) than adjacent non-nest soils (figure 2). However, the increase of N, C and P content in refuse dumps respect to adjacent soils can vary between 40 (*Acromyrmex lobicornis*, Tadey & Farji-Brener 2007) and 100-fold (*Atta sexdens*, Mountinho *et al.* 2003) in Patagonian steppes and Brazilian rain forests, respectively. Despite the above described variability, and to our knowledge, the magnitude of the contribution to soil fertility is among the highest generated by any animal group. Accordingly, nutrient flow is from 16 to 98 times faster in refuse dumps than leaf litter controls (Haines 1978), generating a net positive effect on forest energy balance (Lugo *et al.* 1973).

Table 1. Nutrient content in nest-soil and refuse dump samples from different leaf-cutting ant species. OC = Organic carbon, N = Nitrogen, P = Phosphorous, K = Potassium, Mg = Magnesium, and Ca = Calcium. Note that the unities of nutrient content may change in the different studies. Sources, ant species, sampled substrate, habitat and country are described as follows: (1) Moutinho *et al.* 2003, *Atta sexdens*, refuse samples, rain forest, Brazil. (2) Etter & Botero 1990, *Atta laevigata*, nest soil samples, savanna, Colombia. (3) Farji-Brener & Silva 1995a, *Atta laevigata*, nest soil samples, savanna, Venezuela. (4) Haines 1978, *Atta colombica*, refuse samples, rain forest, Panama. (5) Farji-Brener & Ghermandi 2008, *Acromyrmex lobicornis*, refuse dump samples, Patagonian steppe, Argentina. (6) Tadey & Farji-Brener 2007, *Acromyrmex lobicornis*, refuse dump samples, Monte desert, Argentina. (7) Sousa-Souto *et al.* 2008, *Atta laevigata*, nest soil samples, savanna, Brazil. (8) Farji-Brener unpub.data, *Acromyrmex lobicornis*, refuse dump samples, desert scrubland, Argentina

Source/ Nutrients	OC	N	P	K	Mg	Ca
1	19.5 %	2.6 %	253 μ g/g	603 μ g/g	892 μ g/g	2179 μ g/g
2	1.2 %		3 ppm	0.04 mep/100g		
3	1.8 %	0.1 %	2.4 ppm	0.11 mep/100g	0.2 meq/100g	0.35 meq/100g
4		3.1 %	0.26 %	1.5 %	0.40 %	1.2 %
5	13.8 %	0.8 %	597 ppm	7 mep/100g	6.7 meq/100g	21 meq/100g
6	16 %	0.9 %	350 mg/Kg			
7	6 %		4.5 mg/Kg			0.3 cmol/Kg
8	8 %	0.8 %	65 mg/Kg	2.1 g/Kg		2.5 g/Kg

The quantity of refuse dumps generated by the ants is also surprising. Nests of *Atta sexdens* and *Atta laevigata* in Brazil can contain up to 300 refuse chambers with organic material at 1-6 m depth (Autori 1942, Coutinho 1982). Brazilian nests of *Atta capiguara* can harbor up to 500 Kg of organic waste in underground chambers (Amante 1964, 1967). *Atta colombica* deposited a mean of 9.4 tons/yr of refuse on soil surface, ranging from 34 to 400 kg/yr/colony in a tropical rain forest of Panama (Hertz *et al.* 2007). External refuse dumps of a single nest of *Acromyrmex lobicornis* may cover up to 2 m^2 of soil surface in Patagonian steppes (Farji-Brener & Ghermandi 2000. 2004, 2008). In sum, both the quality and quantity of refuse dumps are a notorious contribution to soil fertility, irrespective of their location (underground chambers or on the soil surface).

Consequences of Soil Enrichment Caused by Leaf-Cutting Ants: Their Effects on Plants

The high concentration of soil nutrients generated by leaf-cutting ants often enhances the growth and performance of plants around the nest area. Several lines of evidence demonstrate that plants can access to the enhanced nutrient content of refuse dumps. First, root abundance is often greater in refuse dumps than in adjacent soils. While subterranean refuse chambers showed higher roots proliferation from adjacent trees (Stahel & Geijskes 1939, Moutinho *et al.* 2003, Sousa-Souto *et al.* 2008), external refuse dumps show higher density of fine roots of adjacent seedlings and herbs (Farji-Brener & Medina 2000, Wirth *et al.* 2003, Farji-Brener & Ghermandi 2004). For example, density of fine roots is 4-folded in the external refuse dumps of *Atta colombica* respect to the adjacent control soils (Haines 1978; Farji-Brener & Medina 2000). Roots of woody plants are 100% higher at depths below 40 cm in nests of *Atta laevigata* than in nearby non-nest soils in Brazilian savannas (Sousa-Souto *et al.* 2008). Second, plants near ant nests often show higher foliar nutrient content than those distant from ant nests (Sousa-Souto *et al.* 2007, Farji-Brener & Ghermandi 2008). Finally, experiments and field measurements of natural isotopes (i. e., ^{15}N) demonstrate that plants assimilate nitrogen from refuse dumps (Sternberg *et al.* 2007, Farji-Brener & Ghermandi 2008). All this evidence strongly support that plants can access to the nutrient content of refuse dumps.

Plants not only gain access to this additional source of nutrients, they also benefit from them. Vegetation associated to ant nests and refuse dumps frequently showed faster growth rate and greater root and leaf biomass when seedlings, and higher number of flowers and seeds when adults. For example, seedlings experimentally planted in refuse dumps showed after few weeks between 1-10 times more foliar area, and leaf and root biomass than seedlings growing on non-nest soils. As a result, adult plants that grew in refuse dumps produced 1-3 times more seeds than non-nest soil plants (Farji-Brener & Ghermandi 2008). Therefore, additional sources of soil nutrients are available for plant growth and reproduction as a product of leaf-cutting ant's activity.

This effect of soil enrichment with consequences on plant performance may affect vegetation dynamics and scale-up to higher spatial scales. Many studies have shown that leaf-cutting ant nests can favor a set of plants different from that of the dominant vegetation. In Paraguayan savannas and Argentinean open woodlands, abandoned or dead *Atta vollenweideri* nests are colonized by woody *Prosopis* spp. The ant nest is a favorable environment for these seedlings because it is free of competing grasses, rich in nutrients, and higher in water content than the surrounding environment (Jonkman 1978, Bucher 1982). In Venezuelan tropical savannas, *Atta laevigata* favors the establishment of several tree species in their nest area (Farji-Brener & Silva 1995b, 1996). In the Brazilian savannas, the exotic grass *Melinis minutiflora* mainly colonizes the open spaces provided by *Atta laevigata* nests (Coutinho 1982). In temperate steppes, the presence of *Acromyrmex lobicornis* nests alters the plant assemblage composition (Farji-Brener & Ghermandi 2000, 2004). Overall, plant abundance, richness and composition are often affected by the presence of leaf-cutting ant nests.

The effect of ant nests on plant assemblages can have repercussions at landscape level. Abandoned or dead nests of *Atta vollenweideri* become 'woody nuclei', accelerating the invasion of woody plants into the Paraguayan savannas and transforming them into open woodlands (Jonkman 1978). Living *Atta laevigata* nests have a synergistic relationship with deciduous tree species in Venezuelan parkland savannas, encouraging the presence of forest groves in an herbaceous matrix (Farji-Brener & Silva 1995b). *Atta* nests increase the availability of space and light levels at soil surface, influencing plant abundance and richness in tropical rain forests (Garretson *et al.* 1998, Farji-Brener & Illes 2000). All this examples illustrate how leaf-cutting ants contribute to soil fertility affecting plant demography, plant assemblages and landscape structure.

Factors that Affect the Contributions of Leaf-Cutting Ants to Soil Fertility

Since the changes in physical and chemical soil properties produced by leaf-cutting ants depend on ant foraging behavior and nest demography, any modification to these two factors will influence the ants' contribution to soil fertility. Temperature affects foraging activities through its effects on oxygen consumption, water loss, and transport cost (Lighton & Feener 1989), which reduces daily and seasonally leaf-cutting ants foraging activity (Mintzer 1979). High temperatures also increase the desiccation rate of the leaf fragments, and well-hydrated leaves are important for fungal growth (Bowers & Porter 1981). In addition, the speed of ant foragers is very slow at temperatures under 15 °C (Mintzer 1979). Therefore, in extreme cold or warm periods ants often do not show foraging activity, reducing their net contribution to soil fertility. Plant abundance and richness is another important factor that affects leaf-cutting ants. High abundance of palatable plants increase ant forage rate and nest development, improving the rate of refuse dump production and nest abundance. Conversely, low abundance of palatable vegetation decreases foraging activity, the production of refuse dumps, ant's reproduction rate, and finally nest abundance (Farji-Brener 2001). Competition for food with other animals also affects the contribution of these ants to soil fertility. For example, in the Monte desert of Argentina, the nutrient levels in refuse dumps showed a strong decrease with an increment of grazing intensity by mammalian herbivores. The drastic decrease in plant cover (from 70% to 20%), concomitant with an increment in grazing intensity, seems to deplete the leaf-cutting ant diet by reducing the chance of encounter between scout ants and palatable plant species. Therefore, the overall reduction of the more palatable plants abundance may cause leafcutters to forage on the remaining less nutrient-rich species. This shift toward less nutritive plant species deprives the nutrient content of refuse dumps, affecting their positive impact to soil fertility (Tadey & Farji-Brener 2007). Finally, frequent burning reduces the positive effects of leaf-cutting ant nests as high-nutrient patches (Coutinho 1984, Sousa-Souto *et al.* 2008). Soil nutrient content in ant nest located in burned habitats could be reduced due to (a) a higher assimilation of nutrients by plants during post-fire recover, (b) a lower foraging activity of the ants (e. g., because of the higher solar irradiation and/or lower vegetation abundance), and (c) a lower activity in underground chambers that may generate the lose of soil nutrients through leaching (Sousa-Souto *et al.* 2008). In sum, the strength of the contribution of leaf-cutting ants to soil fertility is negatively affected by extreme environmental weather, low abundance of palatable vegetation for ants, high abundance of introduced mammalian herbivores and frequent fires.

CONCLUSIONS

Leaf-cutting ants increase soil fertility through two mechanisms. The construction of underground cavities and galleries in their nests increase soil porosity, drainage, and aeration. These changes in soil structure often enhance soil biota activity. On the other hand, large quantities of plant nutrients within the foraging area of a colony are relocated to the nest area as food for the ant's fungus culture (Wirth *et al.* 2003). As a consequence of this process, leafcutters generate a huge quantity of organic matter as waste product. This nutrient-rich organic waste (i. e., refuse dumps) is located in underground chambers or external piles. The above mentioned activities strongly enhance soil fertility; almost all soil nutrients are up to 100 times more concentrated in nest areas (especially in refuse dumps) than in adjacent, non-nest soils. These magnitudes are, to our knowledge, among the highest contribution to soil fertility performed by any animal group. Therefore, the presence of ant nests and their associated refuse dumps affect plant abundance, growth, and reproduction (Farji-Brener & Ghermandi 2004, 2008). The effect of ant nest also impacts on plant assemblages and vegetation landscape structure (Jonkman 1978, Farji-Brener & Silva 1995b). At the ecosystem level, the activity of colonies modifies resource flows directly relocating nutrients and soil particles and indirectly through effects on soil properties (Lugo *et al.* 1978, Haines 1978). Finally, the contribution of leaf-cutting ants to soil fertility ultimately depends on nest demography and foraging activity. Both processes are, in turn, deprived by extreme weather, low availability of palatable plants for leafcutters, high levels of competition with exotic mammalian herbivores, and frequent fires. Particularly, the presence of exotic mammalian herbivores (i. e., via competition for food) and frequent fires (i. e., through reducing plant abundance and ant activities) negatively affects the role of leaf-cutting ants as soil fertilizers. Since the abundance of palatable vegetation for leafcutters, and thus nest density, are higher in disturbed than in more pristine environments (Jaffe & Vilela 1989, Farji-Brener 2001, Vasconcelos & Cherret 1995), the contribution of leaf-cutting ant nests to soil fertility may be more important in early successional stages and disturbed habitats. This fact may have important implications for conservation and restoration. In terms of conservation, ant-nest areas should be especially protected in early successional habitats because they are hot spot of plant productivity and diversity, and may act as a nucleus of plant stablishment, succession and dispersal. From a restoration perspective, the nutrient-rich nest-soils and refuse dumps may be employed as natural, free-access and ecologically sustainable fertilizer to improve soil fertility in degraded areas.

ACKNOWLEDGMENTS

Part of the researches and unpublished data employed in this review was supported by the following grants to AGFB: PIP-CONICET 5110, and FONCyT-PICT 25314, Argentina.

REFERENCES

Alvarado, A., Berish, C., & Peralta, F. (1981). Leaf-cutter ant (*Atta cephalotes*) Influence on the morphology of andepts in Costa Rica. *Soil Science Society American Journal 45*: 790-794.

Amante, E. (1964). Nota previa sobre a estrutura do ninho de una nova formiga sauva (*Atta* sp) (Hymenop., Formicidae). *Biológico 30*: 96-97.

Amante, E. (1967). Prejuizios causados pela formiga sauva en plantacoes de Eucalyptos e Pinus no estado de Sao Pablo. *Silvicultura em Sao Pablo 6*: 355-363.

Autori, M. (1942). Contribucao para o conocimento da sauva (*Atta* spp) 3. Excavacao de un sauverio. (*Atta sexdens*). *Archivos do Instituo Biologico do Sao Pablo 13*: 137-148.

Boulton, A. & Amberman, K. (2006). How ant nests increase soil biota richness and abundance: a field experiment. *Biodiversity and Conservation 15*: 69-82.

Bowers, M. & Porter, S. (1981). Effect on foraging distance on water content of substrates harvested by *Atta colombica*. *Ecology 62*: 273-275.

Bruyn, L. & Conacher, A. J. (1995). Soil modification by termites in the central wheatbelt of Western Australia. *Aust. J. Soil Res. 33*: 179-193.

Bucher, E. (1982). Chaco and Caatinga-South American Arid Savannas, Woodlands and Thickets. In *Ecology of tropical savannas*, ed. Huntley, B. & Walker B. Springer-Verlag, Berlín, pp. 48-79.

Cherrett, J. M. (1989). Leaf-cutting Ants, Biogeographical and Ecological Studies. In *Ecosystem of the world 14b*, *Tropical rain forest ecosystem*, eds. Lieth, H. & Werger, M. Elsevier, New York, pp. 473-488.

Coutinho, L. (1984). Aspectos ecológicos da saúva no cerrado. A saúva, as queimadas e sua possivel relacao na ciclagem de nutrientes minerais. *Boletim Zoologico de la Universidad de Sao Paulo 8*: 1-9.

Coutinho, L. (1982). Aspectos ecológicos da saúva no cerrado. Os murundus de tierra, as características psamofíticas das especies de sua vegetacao e a sua invasao pelo Capim Gordura. *Revista Brasilera de Biología 42*: 147-153.

Cosarinsky, M. & Roces, F. (2007). Neighbor leaf-cutting ants and mound-building termites: comparative nest micro-morphology. *Geoderma 141*: 224-234.

Dangerfield, J., McCarthy, T. & Ellery W. (1998). The mound-building termite *Macrotermes michaelseni* as an ecosystem engineer. *Journal of Tropical Ecology 14*: 507-520.

Etter, A. & J. Botero, P. (1990). La actividad edáfica de hormigas (*Atta laevigata*) y su relación con la dinámica sabana/bosque en los Llanos Orientales (Colombia). *Colombia Amazónica 4*: 77-95.

Farji-Brener, A. G. (2001). Why are leaf-cutting ants more common in early secondary forests than in old-growth tropical forests? An evaluation of the palatable forage hypothesis. *Oikos 92*: 169-177.

Farji-Brener, A. G. & Ghermandi, L. (2008). Leaf-cutting ant nests near roads increase fitness of exotic plant species in natural protected areas. *Proceedings of the Royal Society - Series B. 275*: 1431-1440.

Farji-Brener, A. G. & Ghermandi, L. (2004). Seedling recruitment in the semi-arid Patagonian steppe: facilitative effects of refuse dumps of leaf-cutting ants. *Journal of Vegetation Science 15:* 823-830

Farji-Brener, A. G. & Ghermandi, L. (2000). The influence of nests of leaf-cutting ants on plant species diversity in road verges of northern Patagonia. *Journal of Vegetation Science 11*: 453-460.

Farji-Brener, A. G. & Illes, A. (2000). Do leaf-cutting ant nests make 'bottom up' gaps in neotropical rain forests? A critical review of the evidence. *Ecology Letters 3*: 219-227.

Farji-Brener, A. G. & Medina, C. (2000). The importance of where to dump the refuse: seed banks and fine roots in nests of the leaf-cutting ants *Atta cephalotes* and *Atta colombica. Biotropica 32*: 120-126.

Farji-Brener A.G. & Silva, J. (1995a). Leaf-cutting ants and forest groves in a tropical parkland savanna of Venezuela: facilitated succession? *Journal of Tropical Ecology 11*: 651-669.

Farji-Brener A.G. & Silva, J. (1995b). Leaf-cutting ant nests and soil fertility in a well-drained savanna in western Venezuela. *Biotropica* 27: 250-253.

Fonte, T. S. & Schowalter, T. D. (2005). The influence of a neotropical herbivore (*Lamponius portoricensis*) on nutrient cycling and soil processes. *Oecologia 146*: 423-431.

Forti, L. & Moreira, A. (1997). Architecture of *Atta* nests. In: *Mirmecologia Neotropical.* Anais do VI international pest ant symposium, eds. Delabie J., Campiolo S., Do Nascimiento I., & Ferreira C. Ilheus, Brazil, pp. 91.

Frost, C. & Hunter, M. (2006). Insect Canopy herbivory and frass deposition affect soil nutrient dynamics and export in oak mesocosms. *Ecology 85*: 3335-3347.

Garrettson M., Stetzel, J., Halpern, B., Hearn, J., Lucey, B. & Mckone, M. (1998). Diversity and abundance of understory plants on active and abandoned nest of leaf-cutting ants (*Atta cephalotes*) in Costa Rica rain forest. *Journal of Tropical Ecology 14*:17-26.

Haines, B. (1978). Element and energy flows through colonies of the leaf-cutting ant, *Atta colombica*, in Panama. *Biotropica 10*: 270-277.

Haines, B. (1975). Impact of Leaf-Cutting Ants on Vegetation Development at Barro Colorado Island. In *Tropical ecological systems, trends in terrestrial and aquatic research*, eds. Goley, F. & Medina, E. Springer-Verlag, New York, pp. 99-111.

Herz, H., Beyschlag, W. & Holldobler, B. (2007). Herbivory rate of leaf-cutting ants in a tropical moist forest in Panama at the population and ecosystem scales. *Biotropica 39*: 482-488.

Jaffe, K. & Vilela, E. (1989). On nest densities of the leaf-cutting ant *Atta cepahlotes* in tropical primary forest. *Biotropica 21*: 234-236.

Jonkman, J. (1978). Nests of the leaf-cutting ant *Atta vollenweideri* as accelerators of succession in pastures. *Zeitschrift für Angewandte Entomologie 86*: 25-34.

Lighton, J. & Feener, D. (1989). Water-loss rate and cuticular permeability in foragers of the desert ant Pogonomyrmex rugosus. *Physiological Zoology 62*: 1232-1256.

Lovett, G. M. & Ruesnik, A. R. (1995). Carbon and nitrogen mineralization from decomposing gypsy moth frass. *Oecologia 104*: 133-138.

Lugo, A., Farnworth, E., Pool, D., Jerez, P, & Kaufman, G. (1973). The impact of the leaf cutter ant *Atta colombica* on the energy flow of a tropical wet forest. *Ecology 54*: 1292-1301.

Mintzer, A. (1979). Foraging activity of the Mexican leaf-cutting ant, *Atta Mexicana*, in a Sonoran desert habitat. *Insects Sociaux 26*: 364-372.

Moreira, A., Vitorio, A. & Forti, L. (1995). Arquitetura da colonia da sauva *Atta laevigata* (Hymen., Form.). In: *Anais XII encontro de mirmecologia*. Brazil, pp. 67.

Moutinho, P., Nepstad, D. & Davidson, E. (2003). Influence of leaf-cutting ant nests on secondary forest growth and soil properties in Amazonia. *Ecology 84*: 1265-1276.

Owen, D. (1980). How plants may benefit from the animals that eat them. *Oikos 35*: 230-235.

Salick, J., Herrera, R & Jordan, C. (1983). Termitaria: nutrient patchiness in nutrient-deficient forests. *Biotropica 15*: 1-7.

Schowalter, T. D. (1981). Insect herbivore relationship to the state of the host plant: biotic regulation of ecosystem nutrient cycling through ecological succession. *Oikos 37*: 126-130.

Sousa-Souto, L., Schoereder, J, Schaefer, C. & Silva, W. L. (2008). Ant nests and soil nutrient availability: the negative impact of fire. *Journal of Tropical Ecology 24*: 639-646.

Sousa-Souto, L., Schoereder, J, & Schaefer, C. (2007). Leaf-cutting ants, seasonal burning and nutrient distribution in cerrado vegetation. *Austral Ecology 32*: 758-765.

Stadler, B, Michalzik, B, Muller, T. (1998). Linking aphid ecology with nutrient fluxes in a coniferous forest. *Ecology 79*: 1514-1525.

Stahel, G. & Geijskes, D. (1939). Ueber den bau der nester von *Atta cephalotes* L. and *A. sexdens* L. (Hymenop., Formicidae). *Revista de Entomología 10*: 27-78.

Sternberg, L., Pinzon, M., Moreira, M., Moutinho, P., Rojas, E., & Herre, A. (2007). Plants use macronutrients accumulated in leaf-cutting ant nests. *Proceedings of the Royal Society - Series B 274*: 314-321.

Tadey, M. & Farji-Brener, A. G. (2007). Indirect effects of exotic grazers: livestock decreases the nutrient content of refuse dumps of leaf-cutting ants through vegetation impoverishment. *Journal of Applied Ecology 44*: 1209-1218.

Vasconcelos, H. & Cherrett, M. (1995). Changes in leaf-cutting ant populations (Formicidae; Attini) after the clearing of mature forest in Brazilian Amazonia. *Studies on Neotropical Fauna and Environment 30*: 107-113.

Wagner, D., Brown, M. & Gordon, D. (1997). Harvester ant nests, soil biota and soil chemistry. *Oecologia 112*: 232-236.

Weber, A. (1982). Fungus ants. In: *Social insects*. Vol 4, ed. Hermann H. Academic Press, London, pp. 255-363.

Wirth, R., Herz, H., Rye, l., Beyschlag, W. & Holldobler, B. (2003). *Herbivory of Leaf-Cutting Ants*. Springer-Verlag, Berlin.

In: Soil Fertility
Editors: D. P. Lucero and J. E. Boggs

ISBN: 978-1-60741-466-93

Chapter 7

SOIL MICROBIAL BIOMASS AND SUSTAINABLE FARMING —ENHANCING NUTRIENT RETENTION IN SOILS THROUGH MANAGEMENT OF MICROBIAL BIOMASS

Song Qiu, Richard Bell and Arthur McComb
School of Environmental Science, Murdoch University,
Murdoch, WA 6150, Australia

ABSTRACT

Soil fertility is critical in sustainable farming and needs to be considered not only for crop productivity but also for the protection of aquatic environments. Evidence suggests that soil microbial biomass is a flexible nutrient store, and with appropriate management such a nutrient store could have potential dual benefits for improved crop productivity and enhanced nutrient retention in soils to minimize nutrient export from agricultural land to environmental waters. To date, most research has focused on stimulating the release of nutrients from the microbial biomass pool for crop uptake. By contrast, our understanding of microbial biomass as a soil regulator to buffer nutrient supply and potential leaching loss is still limited. Such a role however could be usefully explored in heavily weathered and nutrient-deficient sandy soils under Mediterranean climate conditions, such as soils in the coastal region of southwestern Australia. Here we present our views on the critical aspects of microbial immobilisation in relation to soil fertility and nutrient retention, and propose a conceptual framework for such a role of microbial nutrient retention. This is supported by previously reported data on microbial immobilisation and nutrient retention. The critical issues such as the size of soil microbial biomass and the time scale for biomass turnover, as well as its potential mobility in soil matrices are discussed along with the research needs in understanding the role of soil microbial processes in sustained soil fertility and environmental conservation.

INTRODUCTION

The environmental concerns about agricultural land use over the past few decades were particularly associated with the overuse of fertiliser, which leads to the export of nutrients from farm land to downstream water and causes eutrophication and deterioration of water quality. In southwestern Australia, nutrient leaching is also attributable to the poor nutrient sorption by the sandy soils prevailing on the coastal region, which have little capacity to retain nutrients especially phosphorus and nitrogen (McArthur 1991). Managing the leaching loss from fertilised farm land on such sandy soils is a challenging task for maintaining soil quality and sustainable farming, which requires an holistic and integrated approach that tackles the problem through both reducing input (fertilizers) and increasing soil nutrient retention.

Soils are complex systems and nutrient transfers and transformations within them can be multi-level and hierarchical, but basically they are biotic and abiotic. Abiotic processes such as phosphorus (P) adsorption-desorption control the concentration of inorganic P (P_i) and other ions in soil solution. Biotic processes such as heterotrophic microbial activity mediate organic turnover and hence the transfers of P_i and soluble inorganic nitrogen (N_i) between organic and inorganic pools. From the 'nutrient leaching' perspective, it is critical to maintain low soluble P_i and N_i in fertilised soils because the leaching, when it happens, is often associated with higher nutrient concentration in soil solution, especially when the concentration exceeds a threshold, the so called 'critical change point' (Hesketh and Brookes 2000).

In this paper, we discuss the potential role of microbial nutrient retention in sustainable nutrient management. Microbial immobilisation and biomass as a nutrient store is explored in the context of nutrient conservation and amelioration of soil nutrient leaching. Enhancing nutrient retention in soil through adequately managing microbial immobilisation is assessed from the perspective of sustainable farming, as well as for the protection of environmental water quality.

POTENTIAL OF MICROBIAL NUTRIENT RETENTION

Microbial biomass is ubiquitous in soils. It is a critical component in nutrient cycling, energy flow and ecosystem functioning (Ohtonen et al. 1997; Richardson 2001). On a global scale, the distribution of soil microbial biomass is correlated with latitude, soil carbon and pH (Wardle 1998). At a regional scale, the abundance of soil microbial biomass is associated with soil carbon, soil texture and soil microclimate conditions (Murphy et al. 1998; Grierson and Adams 2000; Mendham et al. 2002).

Under the Mediterranean climate of southwestern Australia, nutrient leaching occurs during periods of concentrated rain particularly when the soils are already wet. The high leaching 'risk' is typically associated with high nutrient concentration in soils particularly those with a predominantly sandy matrix. These conditions may, however, favour microbial biomass development and potentially the formation of the 'biofilm' system, provided that other limiting factors such as carbon, nutrients and low temperature do not become major constraints or can be adequately managed.

From a nutrient management perspective, it is important to know whether microbial biomass can play a role to maintain low P_i and N_i concentration in soil solution so that the leaching loss and export of nutrients to surface and ground water can be reduced. Can such biomass potential be maximised so that biomass acts as an active and flexible store for the purpose of nutrient conservation and sustainable agriculture? We believe that there is unexploited potential for soil microbial biomass to take on such a role. We propose a concept of promoting 'microbial retention' in soils to offset the low 'abiotic sorption' and illustrate it by reference to sandy soils of regional south-western Australian conditions (Figure 1). A number of the critical aspects of the concept are addressed below.

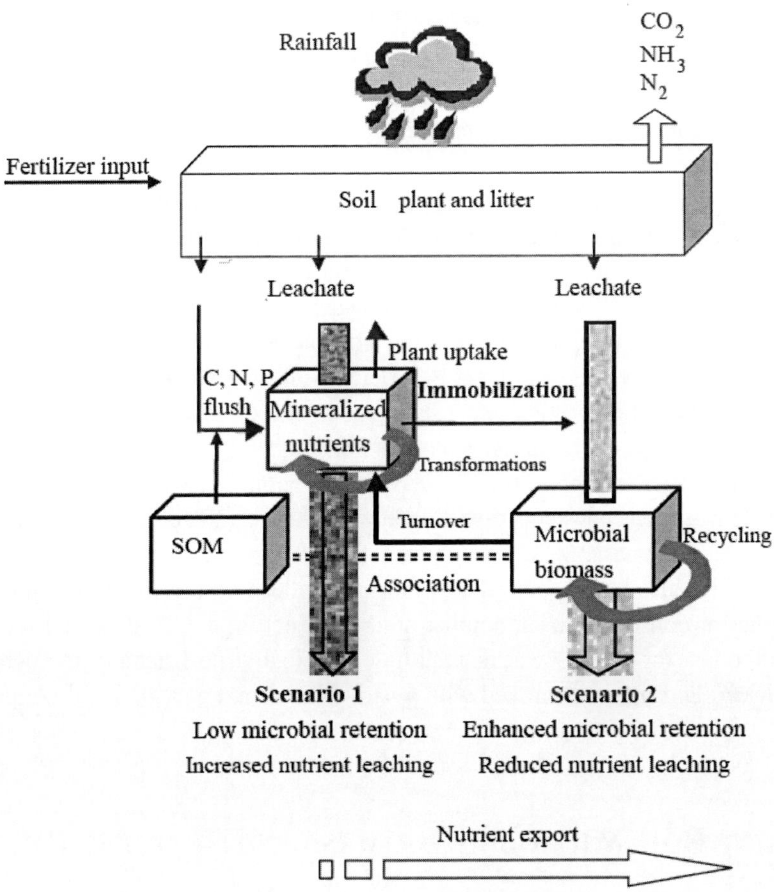

Figure 1. Schema of soil nutrient stores (boxes), nutrient fluxes and leaching pathways relative to microbial immobilisation and retention.

Firstly, soil microbial biomass has a large capacity to retain and store labile nutrients (Frossard et al. 2000). For example, microbial biomass has been reported to retain 17-290 kg P ha^{-1} in agricultural ecosystems, more than the P taken up annually in herbage (Brookes et al. 1984; He et al. 1997). Such capacity, often referred to as 'net immobilisation', occurs under appropriate substrate and nutrient conditions. Thus significant potential can be anticipated in the use of soil microbial biomass to maintain low P_i and N_i in soil solution during the leaching period following rainfall.

Secondly, microbial biomass tends to maximize its metabolic potential under favorable conditions. Such a biomass capacity has been shown to play a significant role in aquatic systems, for example, the rapid depletion of water column P or N during cynobacterial blooms (caused by immobilisation to cynobacterial biomass), which resulted in immobilization of about 80 % of nutrients retained in lakes and recycled within the system (Harris 1999). Such evidence can also be observed from the microbial biofilm processes developed for nutrient removal in waste-water plants (Fitch et al. 1999), as well as in artificial wetlands used for waste water treatment and nutrient removal (Tate III 2000).

An impressive microbial capacity for retaining labile nutrients was observed in a woodland catchment of local wetlands of southwestern Australia. We reported that surface litter from the catchment could release significant amounts of P (2-5 mg P L^{-1}) in leachate during the 'first flush' following the first heavy rains of the season. A significant pool of the leached P was retained by microbial biomass. Our data suggest that most 'leachable P' in soils (anion-exchange-membrane extractable P) was immobilised by microbial biomass over the early wet season (Qiu et al. 2002; 2003). There was a three-fold increase in biomass-P along a catchment-wetland transect after the early rains, resulting in only about 5 % of 'leachable P' remaining in soils compared with that held in soil before the rains. The large immobilisation was attributable to the increased soil moisture as well as nutrient acquisition by microorganisms from leached surface litter. Such an observation was not alone, it was consistent with previously reported data in the tropical region by Oberson et al. (2001), who suggested that in severely weathered and P-deficient soils, the role of biological processes can override physiochemical P sorption in mediating soil solution P.

The large capacity of biomass retention was also reported on soil N. In Western Australia, a much greater proportion of the N can be immobilised in the microbial biomass than in more temperate ecosystems. Sparling et al. (1995) estimated that 31-73 kg N ha^{-1} was held by microbial biomass, which comprised 8-15 % of the total soil N. The microbial immobilisation has also been reported as a mechanism leading towards reduced leaching and improved nutrient conservation. In another study, the net immobilisation and retention of N was reported to last for over 29 weeks, resulting in virtually 'no-leaching' in microlysimeters when *Eucalyptus* leaf litter was mixed with a native forest and a pasture soil (Aggangan et al. 1999).

MANAGING SOIL MICROBIAL BIOMASS AND NUTRIENT RETENTION

Microbial processes of immobilisation and mineralisation are often associated with soil 'priming effects', a term for the characteristic, short-term but strong changes in nutrient flow and turnover following moderate soil treatments (Dalal 1998; Kuzyakov et al. 2000). The 'priming effects' may either be viewed as 'positive' or 'negative' in relation to organic mineralisation. A 'negative' effect produces net immobilisation of nutrients and slow organic turnover (Magid et al. 1999; Kuzyakov et al. 2000), which is of interest for soil nutrient retention.

Classes of different organic substrate appear to trigger different priming effects in relation to carbon. For example, the amino acids, glutamate, and aspartate can cause positive priming effects, while glucose, cellulose, wheat straw, and sewage sludge cause negative priming effects (Dalenberg and Jager 1989). The priming effects have been reported not only for carbon but also on nitrogen and phosphorus. These effects can be triggered not only by available organic C, but also by addition of a limiting nutrient, or even by a change in soil physical condition such as drying and rewetting (Kuzyakov et al. 2000). Dry-rewet treatment has been reported to stimulate soil nutrient cycling due to osmoregulatory shock, altered organomineral surfaces, soil aggregation and increased microbial turnover. Thus drying and rewetting rapidly increased the N immobilisation by 3-fold in the forest floor, though over time N mineralisation may gradually become dominant over immobilization (Pulleman and Tieteman 1999). Under dry conditions, biodegradable nutrients may accumulate because of insufficient water to support microbial activity. Upon rewetting (rain), rapid microbial growth may occur, owing to a flush of nutrients from microbial biomass and decomposable organic substrates accumulated during the drying period (Sparling and Ross 1988; McNeill et al. 1998; Murphy et al. 1998).

Questions remain concerning the direction of microbial activity following the initial biomass immobilisation. In some circumstances microbial activity results in net immobilisation of nutrients, in others net mineralisation. Such microbial directional behaviors appear to be associated with substrate or nutrient availability. As a rule of thumb, if more P is available in soil organic matter (SOM) than is needed for the rate of C assimilation, net mineralisation of P occurs. Conversely, if insufficient P is available from SOM for C assimilation, bioavailable P in soil solution must be used, which leads to net immobilisation (Mullen 1998). This rule also applies to microbial immobilisation for N (Tate 2000), S (Chapman 1997), and presumably other organically-bound nutrients. Hence, mineralisation-immobilisation may be considered a kinetic response of microbial biomass to nutrient and substrate availability, and potentially manageable through manipulating these conditions.

Knowledge about the substrate- and nutrient- dependency of microbial mineralisation and immobilisation behavior appears to be useful for understanding some published field and laboratory results. For example, a study showed that an increased C substrate supply in S-deficient soils could cause S immobilisation and induce S limitation of microbial biomass, thereby depressing C mineralisation (Chapman 1997). On the other hand, immobilisation of N in forest soils is commonly observed, which can be attributed to higher C/N ratios of soil organic matter (SOM) from forest litter; therefore the net mineralisation in soil can be retarded typically by 12-24 months after litter application (Kirschbaum and Paul 2002). Elsewhere wheat straw, wheat stubble and sheep manure, organic materials that usually have higher C/N ratios, were reported to induce rapid N immobilisation, so no net N-mineralisation was detected until day 52 (Corbeels et al. 1999). In another study, cellulose amendments (high C content) produced a significant reduction of ammonia volatilization in a calcareous sandy soil due to increased microbial immobilisation (He et al. 2002). Poudel et al. (2002) compared soil N availability and leaching potential between organic, low-input and conventional farming systems, and found significantly lower N mineralization rates and leaching potential in organic and low-input farming systems. Moreover, frequent watering (enhanced microbial activity and reduced water stress) was reported to cause net immobilisation of N and P in pine forest litter (Tuininga et al. 2002).

Further quantitative information appears to be critical to increase the predictability of the effects of carbon to nutrient ratios. To this end, we used a wide range of C/N (1-167) and C/P (7-651) ratios in two sandy soils of southwestern Australia, to test microbial response to the amendments in relation to N immobilisation and mineralisation behavior. Inorganic N in soil water was measured after amended incubation, and net immobilisation or mineralisation estimated as the difference in soil water N before and after incubation. The results showed that all treatments with additional C-supply resulted in net N immobilisation, while the "non-C" treatments with only nutrients (N, P and N+P) led to net N-mineralisation. C/N ratio appears more critical than C/P in directing microbial N immobilisation or mineralization in these sandy soils. Hence N remaining in soil water decreased exponentially with increasing initial C/N. It is suggested that the immediate availability of C, N, and P and their relative supply compared to that of the cell physiological demands underlay microbial responses, and in the short term, determine the direction of the two opposing microbial processes–N mineralisation and immobilisation (Qiu et al. 2008).

Related Issues, Uncertainty and Further Research

Biomass Turnover

Microbial biomass is usually considered as bioavailable to plants through turnover, which returns a proportion of plant available nutrients to the soil solution. Therefore increased microbial biomass and microbial nutrients can be viewed as a source of nutrients (inactivated) available for crop needs. This process may, however, be deflected by recycling of C and nutrients within the microbial biomass through at least two pathways: an intracellular turnover of metabolites for biosynthesis and cell maintenance, and; the immediately reuse of mineralised products by microorganisms. It is still not clear how much of the nutrients would likely to be recycled within the biomass compartment and how much could be available to plants. More quantitative information about the competing processes and the partitioning of these nutrient compartments is needed in order to better understand the fate of soil biomass nutrients

While the main strategy here is to maximise the retention of nutrients following enhanced immobilisation, which temporarily inactivates mineralisation while maintaining microbial vitality over the risk period, presumably from weeks to months during periods when there is highest risk of nutrient leaching. Overall this 'inactivation' of mineralisation is transient and may coincide with periods of low plant growth, so it is not expected to affect overall mineralisation and nutrient availability to plants in such systems.

For an increased protection of leaching loss via maintaining low levels of nutrients in soil solution, soil microbial biomass may need to have a 'longer-term role' in nutrient retention and this may be achieved by appropriate soil management. Microbial biomass is reported to survive in soil for many months without substrate supply (Joergensen et al. 1990). Under substrate-deficient conditions, biomass may enter a 'resting' phase, while maintaining 'metabolic alertness' for future substrate supply. During its resting phase, biomass maintains protein-synthesizing capacity, such as adenosine 5'-triphosphate (ATP) and high-energy charge as a survival strategy (De Nobili et al. 2001).

A range of turnover times for microbial biomass, varying with soil and the methodology, have been reported. These include 45 –215 days (Sakamoto and Hodono 2000) and 82-95 days (Kouno et al. 2002) for microbial C; and 2.5 years (Brookes et al. 1984), 37 and 42 days (Kouno et al. 2002) and 130-215 days (Chen and He 2002) for microbial P. In contrast, microbial N turnover is longer, in the ranges of 793-923, 747-912, and 706-1410 days (Nira 2000). Overall, the reported times for microbial biomass turnover appear of limited use in predicting potential nutrient retention due to their soil and substrate dependency. The turnover time appears dependent on substrate and nutrient conditions, as well as soil temperature and moisture. It is soil specific, related to texture, CEC, macroaggregates, and appears to be reflected in the amount of biomass C in soil (Sakamoto and Hodono 2000). Under the seasonal Mediterranean climate, soil moisture can be a major factor controlling microbial activity and turnover soils (Sparling et al. 1995). The biomass turnover time may also be manageable. De Nobili et al. (2001) reported that that microbial activity can be prolonged through intermittent application of an appropriate trigger such as a small amount of bioavailable carbon, an effect attributed to direct stimulation of a biomass response rather than microbial mineralisation of SOM.

It should be noted that microbial biomass includes a variety of species having different life-spans and physiology including C/N and C/P ratios. There is a potential turnover pathway through the feeding relationships of microfauna and mesofauna, but studies showed the role of mesofauna in mineralisation of nutrients to be fairly small (Hassink et al. 1994; Liiri et al. 2002).

Mobility and Leachability

Soil microbial biomass commonly includes a significant pool of large, filamentous fungal hyphae, which are components of microbial-aggregates and biofilms, i.e. microbial cells in an adhesive, polysaccharide material formed at the interface between soil particles and water (Tate 2000). Microbially-bound nutrients are in particulate, protected forms and are often associated with soil organic matter and soil aggregates, therefore the microbial pool of nutrients is considered less mobile in soil matrix, and the transport of microbial nutrients via leaching is expected to be more manageable than their original soluble forms.

Under an Australian seasonal climate, most nutrient loads would occur during 'storm' events in the wet season, which is reflected in a sudden increase in drainage volume and concentration (Croke and Jakeman 2001). Overland flow may be generated when rainfall intensity exceeds the infiltration capacity of soil matrix (Heathwaite 1997), but such transport pathways appear to be limited on sandy landscapes, such as the soils on the coastal sand dune systems in southwestern Australia (Townley et al. 1993). Subsurface pathways transport nutrients either through the soil matrix and/or by preferential flow through macropores (Simard et al. 2000). The role of these pathways in transporting microbial materials is largely unknown, but whether it is overland flow or preferential flow, a large portion of flow would bypass the soil matrix, reducing the opportunity to flush out microbially-bound nutrients occluded within the soil matrix and aggregates.

Overall, under a mediterranean climate we may expect a rapid increase in moisture and nutrient flux during early seasonal rains, exacerbated by leaching of surface litter and nutrients accumulated during the dry season, leading to stimulation of soil microbial activity. Soil microbial activity may be negligible under dry summer conditions in the region, but will quickly recover after rewetting. Initially, the increased moisture and nutrient availability will enhance microbial uptake and immobilisation of available nutrients. The stimulative response and the level of immobilisation though would vary by soils, being dependent on soil texture, nutrient and carbon availability, as well as microclimate conditions. The initial period of immobilisation can be followed by increased mineralisation if there is low C/P or C/N in SOM, or vice versa. To enhance and maintain the period of net immobilisation and retention it may be necessary to subsequently intervene through soil management. Interventions should focus on applying organic matter of higher C/N and N/P ratios to increase carbon availability, and soil management to improve microbial habitat to facilitate biomass uptake as well as retarding nutrient mineralisation. Further research is anticipated to provide a critical understanding of the management techniques for enhancing microbial retention, to achieve the goals of reduced nutrient export while sustaining soil fertility and protecting environmental water quality.

REFERENCES

Aggangan, RT; O'Connell, AM; McGrath, JF and Dell, B. 1999. The effects of *Eucalyptus globulus* Labill. leaf litter on C and N mineralization in soils from pasture and native forest. *Soil Biology & Biochemistry.* 31, 1481-1487.

Brookes, PC; Powlson, DS and Jenkinson, DS. 1984. Phosphorus in the soil microbial biomass. *Soil Biology & Biochemistry.* 16, 169-175.

Chapman, SJ. 1997. Carbon substrate mineralization and sulphur limitation. *Soil Biology & Biochemistry.* 29, 115–122.

Chen, GC and He, ZL. 2002. Microbial biomass phosphorus turnover in variable-charge soils in China. *Communications in Soil Science & Plant Analysis.* 33, 2101-2117.

Corbeels, M; Hofman, G and Van Cleemput, O. 1999. Simulation of net N immobilisation and mineralisation in substrate-amended soils by the NCSOIL computer model. *Biology & Fertility of Soils.* 28, 422-430.

Croke, BFW and Jakeman, AJ. 2001. Prediction in catchment hydrology: an Australian perspective. *Marine & Freshwater Research.* 52, 65-79.

Dalal, RC. 1998. Soil microbial biomass -what do the numbers really mean? *Australian Journal of Experimental Agriculture.* 38, 649-665.

Dalenberg, JW and Jager, G. 1989. Priming effect of some organic additions to ^{14}C-labeled soil. *Soil Biology & Biochemistry.* 21, 443-448.

De Nobili, M; Contin, M; Mondini, C. and Brookes, PC. 2001. Soil microbial biomass is triggered into activity by trace amounts of substrate. *Soil Biology & Biochemistry.* 33, 1163-1170.

Fitch, MW; Murphy, JB and Sowell, SS. 1999. Biological fixed-film systems. *Water Environment Research.* 71, 638-656.

Frossard, E; Condron, L M; Oberson, A; Sinaj, S and Fardeau, JC. 2000. Processes governing phosphorus availability in temperate soils *Journal of Environmental Quality.* 29, 15-23.

Grierson, PF and Adams, MA. (2000). Plant species affect acid phosphatase, ergosterol and microbial P in a Jarrah (*Eucalyptus marginata* Donn ex Sm.) forest in south-western Australia. *Soil Biology & Biochemistry.* 32, 1817-27.

Harris, GP. 1999. This is not the end of limnology (or science): the world may well be a lot simpler than we think. *Freshwater Biology.* 42, 689-706.

Hassink, J; Neutel, AM and De Ruiter, PC. 1994. C and N mineralization in sandy and loamy grassland soils: the role of microbes and microfauna. *Soil Biology & Biochemistry.* 26, 1565-71.

He, ZL; Calvert, DV; Alva AK; Li, YC and Banks, DJ. 2002. Clinoptilolite zeolite and cellulose amendments to reduce ammonia volatilization in a calcareous sandy soil. *Plant and Soil.* 247, 253-260.

He, ZL; Wu, J; O'Donnell, AG and Syers, JK. 1997. Seasonal responses in microbial biomass carbon, phosphorus, and sulphur in soils under pasture. *Biology & Fertility of Soils.* 24, 421-428.

Heathwaite, AL. 1997. Sources and pathways of phosphorus loss from agriculture. p. 205-223. In H. Tunney et al. (ed.) Phosphorus loss from soil to water. CAB Int., Wallingford, UK.

Hesketh, N and Brookes, PC. 2000. Development of an indicator for risk of phosphorus leaching. *Journal of Environmental Quality, 29, 105-110.*

Joergensen, RG; Brookes, PC and Jenkinson, DS. 1990. Survival of soil microbial biomass at elevated temperatures. *Soil Biology & Biochemistry.* 22, 1129-1136.

Kirschbaum, MUF and Paul, KI. 2002. Modelling C and N dynamics in forest soils with modified version of the CENTURY model. *Soil Biology & Biochemistry.* 34, 341-354.

Kouno, K; Wu, J and Brookes, PC. 2002. Turnover of biomass C and P in soil following incorporation of glucose or ryegrass. *Soil Biology & Biochemistry.* 34, 617-622

Kuzyakov, Y; Friedel, JK and Stahr, K. 2000. Review of mechanisms and quantification of priming effects. *Soil Biology & Biochemistry.* 32, 1485-1498.

Liiri, M; Setala, H; Haimi, J; Pennanen, T and Fritze, H. 2002. Soil processes are not influenced by the functional complexity of soil decomposer food webs under disturbance. *Soil Biology & Biochemistry.* 34, 1009-1020.

Magid, J; Kjaergaard, C; Gorissen, A and Kuikman, PJ. 1999. Drying and rewetting of a loamy sand soil did not increase the turnover of native organic matter, but retarded the decomposition of added C-14-labelled plant material. *Soil Biology & Biochemistry.* 31, 595–602.

McArthur, WM. 1991. Reference soils of south-western Australia. Department of Agriculture, Western Australia. Perth: Australia. 265 pp.

McNeill, AM; Sparling, GP; Murphy, DV; Braunberger, P. and Fillery, IRP. 1998. Changes in extractable and microbial C, N, and P in a Western Australian wheatbelt soil following simulated summer rainfall. *Australian Journal of Soil Research.* 36, 841-854.

Mendham, DS; O'Connell, AM and Grove, TS. 2002. Organic matter characteristics under native forest, long-term pasture, and recent conversion to Eucalyptus plantations in Western Australia: microbial biomass, soil respiration, and permanganate oxidation. *Australian Journal of Soil Research.* 40, 859-872.

Mullen, MD. 1998. Transformation of other elements. In DM Sylvia, J.J Fuhrmann, PG Hartel and DA Zuberer Eds. Principles and applications of soil microbiology. Prentice-Hall: New Jersey.

Murphy, DV; Sparling, GP; Fillery, IRP; McNeill, AM and Braunberger, P. 1998. Mineralisation of soil organic nitrogen and microbial respiration after simulated summer rainfall events in an agricultural soil. *Australian Journal of Soil Research.* 36, 231-246.

Nira, R. 2000. Comparison between flux of nitrogen through microbial biomass and estimated amount of nitrogen mineralization by kinetic method in upland soil. Nippon Dojo Hiryogaku Zasshi 71, 388-390 (From Chemical abstracts: CAN 133:321397).

Oberson, A; Friesen, DK; Rao, I M; Buhler, S and Frossard, E. 2001. Phosphorus transformations in an Oxisol under contrasting land-use systems: the role of the soil microbial biomass. *Plant & Soil.* 237, 197-210.

Ohtonen, R; Aikio, S and Väre, H. 1997. Ecological theories in soil biology. *Soil Biology & Biochemistry.* 11/12, 1613-1619.

Poudel, DD; Horwath, WR; Lanini, WT; Temple, SR and van Bruggen, AHC. 2002. Comparison of soil N availability and leaching potential, crop yields and weeds in organic, low-input and conventional farming systems in northern California. *Agriculture, Ecosystems & Environment.* 90, 125-137.

Pulleman, M and Tietema, A. 1999. Microbial C and N transformations during drying and rewetting of coniferous forest floor material. *Soil Biology & Biochemistry.* 31, 275–285.

Qiu, S; McComb, AJ and Bell, RW. 2002. Phosphorus leaching from litterfall in wetland catchments of the Swan Coastal Plain, southwestern Australia. *Hydrobiologia.* 472, 95-105.

Qiu, S; McComb, AJ and Bell, RW. 2008. Ratios of C, N and P in soil water direct microbial immobilisation–mineralisation and N availability in nutrient amended sandy soils in southwestern Australia. *Agriculture, Ecosystems and Environment* 127, 93–99.

Qiu, S; McComb, AJ; Bell, RW and Davis, JA. 2003. Nutrient response to soil and litter metabolic activity in a transect across a seasonal wetland. *Marine & Freshwater Research.* 54, 243-252.

Richardson, AE. 2001. Prospects for using soil microorganisms to improve the acquisition of phosphorus by plants. *Australian Journal of Plant Physiology.* 28, 897-906.

Sakamoto, K. and Hodono, N. 2000. Turnover time of microbial biomass carbon in Japanese upland soils with different textures. *Soil Science & Plant Nutrition.* 46, 483-490.

Simard, RR; Beauchemin, S and Haygarth, PM. 2000. Potential for preferential pathways of phosphorus transport. *Journal of Environmental Quality. 29, 97-105.*

Sparling, GP and Ross, DJ. 1988. Microbial contribution to the increased nitrogen mineralization after air-drying of soils. *Plant & Soil.* 105, 163-167.

Sparling, GP; Murphy, DV; Thompson, RD and Fillery, IRP. 1995. Short-term net N mineralisation from plant residues and gross and net mineralisation from soil organic N after rewetting of a seasonally-dry soil. *Australian Journal of Soil Research.* 33, 961-973.

Tate III, RL. 2000. *Soil microbiology.* John Wiley & Sons, Inc.: New York.

Townley, LR; Turner, JV; Barr, AD; Trefry, MG; Wright, KD; Gailitis, V; Harris, GJ and Johnston, CD. (1993) Wetlands of the Swan Coastal Plain, Volume 3, Interaction between lakes, wetlands and aquifers. Water Authority of Western Australia. 115 pp.

Tuininga, AR; Dighton, J and Gray, DM. 2002. Burning, watering, litter quality and time effects on N, P, and K uptake by pitch pine (*Pinus rigida*) seedlings in a greenhouse study. *Soil Biology & Biochemistry.* 34, 865-873.

Wardle, DA. 1998. Controls of temporal variability of soil microbial biomass: a global-scale synthesis. *Soil Biology & Biochemistry.* 30, 1627-1637.

In: Soil Fertility
Editors: D. P. Lucero and J. E. Boggs
ISBN: 978-1-60741-466-7

Chapter 8

CAN SOIL FERTILITY BE MODELLED?

B. Foereid
Plant & Soil Science, Cruickshank Building, School of Biological Science,
University of Aberdeen, King's College, UK

Several simulation models for carbon and nitrogen turnover exist (e.g Parton et al., 1987; Jenkinson et al., 1987; Williams, 1990; Hansen et al., 1993; Bradbury et al., 1993). Most of them are mainly for agricultural systems, but some can deal with grasslands and forests as well. The interest in model development has been driven first by an interest in predicting nitrogen leaching and later by an interest in green house gas emissions and soil carbon storage. Modelling nitrogen availability and leaching when large and known quantities of nitrogen fertilisers are added is relatively easy. It is much more difficult to predict availability of nitrogen and other nutrients under natural conditions and low input agriculture. The problem is that under such conditions, predicting nutrient levels at any one time depend on accurate prediction of mineralisation and immobilisation rates. These are temporally and spatially variable, and the factors that control them are only partially known (e.g. Vinten et al., 2002; Silva et al., 2005). Some measurable soil parameters are known to be related to soil fertility, such as carbon and clay content (Sylvia et al., 1998). However, soil fertility can not be fully predicted from measurable parameters, so the concept of soil fertility is not fully understood. Soil classification systems (for example FAO classification) usually attach a measure of suitability for agriculture to each soil type, but this is based on experience rather than an understanding of how the fertility comes about. Process based models can only describe processes that are understood, and therefore struggle to describe natural soil fertility adequately. An alternative approach is to introduce soil fertility as calibrated factor based on soil type in the models (Fisher et al., 2002).This can be a good way to get relatively accurate predictions of yield, but does not really improve our understanding of the concept of soil fertility. Furthermore, when also other parameters such as green house gas emission need to be predicted, a more quantitative understanding of the concept of soil fertility is urgently needed. This will require both a better understanding of the soil itself and the interaction between soil and inputs of fertilisers and organic material.

REFERENCES

Bradbury,N.J., Whitmore,A.P., Hart,P.B.S., Jenkindon,D.S.. Modelling the fate of nitrogen in crop and soil in the years following application of ^{15}N-labelled fertilizer to winter wheat. *Journal of Agricultural Science.* 121, 363-379. 1993.

Fisher,G., van Velthuizen,H., Shah,M., Nachtergaele,F.O., 2002. Global Agro-ecological Assessment for Agriculture in the 21st Century: *Methodology and Results.* IIASA, Vienna, 1-117pp.

Hansen,S., Jensen,N.E., Nielsen,N.E., Svendsen,H., 1993. Description of the Soil Plant System Model Daisy. Basic principles and modelling approach. Jordbrugsforlaget, The Royal Veterinary and Agricultural University, Copenhagen.

Jenkinson,D.S., Hart,P.B.S., Rayner,J.H., Parry,L.C., 1987. Modelling the turnover of organic matter in long-term experiments at Rothamsted. *INTECOL Bulletin.* 15 1-8.

Parton,W.J., Schimel,D.S., Cole,C.V., Ojima,D.S., 1987. Analysis of Factors Controlling Soil Organic Matter Levels in Great Plain Grasslands. *Soil Science of America Journal.* 51, 1173-1179.

Silva,R.G., Jorgensen,E.E., Holub,S.M., Gonsoulin,M.E., 2005. Relationships between culturable soil microbial populations and gross nitrogen transformation processes in a clay loam soil across ecosystems. *Nutrient Cycling in Agroecosystems.* 71, 259-270.

Sylvia,D.M., Fuhrmann,J.J., Hartel,P.G., Zuberer,D.A., 1998. *Principles and Applications of Soil Microbiology.* Prentice Hall, New Jersey.

Vinten,A., Whitmore, Whitmore,A., Bloem,J., Howard, Howard,R., Wright, Wright,F., 2002. Factors affecting N immobilisation/mineralisation kinetics for cellulose-, glucose- and straw-amended sandy soils. *Biology and Fertility of Soils.* 36, 190-199.

Williams,J.R., 1990. The erosion productivity impact calculator (EPIC) model: a case history. *Philosophical Transactions of the Royal Society of London.* 329, 421-428.

In: Soil Fertility
Editors: D. P. Lucero and J. E. Boggs
ISBN: 978-1-60741-466-7

Chapter 9

SOIL CHARCOAL AMENDMENTS MAINTAIN SOIL FERTILITY AND ESTABLISH A CARBON SINK – RESEARCH AND PROSPECTS

Christoph Steiner[1]
Biorefining and Carbon Cycling Program, Driftmier Engineering Center, The University of Georgia, Athens, GA 30602, USA, e-mail: christoph.steiner@biochar.org

CHARCOAL AS SOIL AMENDMENT

Sustaining soil fertility is a major agricultural constraint in the Amazon Basin (Tiessen, et al., 1994), thus shifting cultivation accompanied with slash and burn agriculture is the prevailing agricultural practice in the humid tropics.

In addition to the predominant and unproductive Ferralsols and Acrilols, an exceptional dark soil is well known by the indigenous people and colonists for its sustained soil fertility. According to its dark color and origin ,the soil was termed *Terra Preta de Indio*. Smith (1879) and Katzer (1903) were among the first who described the *Terra Preta's* properties and presumed its cultural origin. Later their assumptions were strengthened by the soil scientists Sombroek (1966) and Zech, et al. (1990). *Terra Preta* contains significantly more carbon (C), nitrogen (N), calcium (Ca), and phosphorus (P), and the cation exchange capacity (CEC), pH value, and base saturation are significantly higher in *Terra Preta* soils than in the surrounding Ferralsols and Acrisols (Zech, et al., 1990; Glaser, et al., 2000). *Terra Preta* soils contain up to 70 times more black C than the adjacent soils. Due to its polycyclic aromatic structure, black C is chemically and microbially stable and persists in the environment over centuries (Glaser, et al., 2001b). Today, the anthropic origin of *Terra Preta de Indio* is generally accepted and its fertility is most likely linked to an accumulation of P and Ca, associated with bone apatite (Zech, et al., 1990; Lima, et al., 2002) and black C as charcoal (Glaser, et al., 2001a). The evidence that *Terra Preta* was manmade, and thus the proven feasibility to transform one of the most infertile soils into one of the most productive soil inspired charcoal research.

Slash and Char was described as an alternative to slash and burn (Steiner, 2007) and (Steiner, et al., 2004b) observed that charcoal is currently used by Amazonian settlers to improve soil fertility. If a forest is burned, only around 2-3% of the above-ground C is converted into charcoal (Fearnside, et al., 2001), but charcoal production can capture 50% of the above-ground C. If the charcoal is not used as fuel (e. g., soil amendment) it has a very high recalcitrance against biological or chemical decay and stores the carbon over centuries or millennia. *Slash and char* is an alternative agricultural method producing charcoal out of the aboveground biomass instead of converting it to CO_2 through burning. If re-growing resources are used, *slash and char* could establish as a significant carbon sink and could be an important step towards sustainability and SOM conservation.

On a global scale, the total carbon release flux due to fire is of the order of 4-7 Pg of carbon per year. This flux is almost as large as the rate of fossil fuel consumption (about 6 Pg per year in 1990) (Goudriaan, 1995). Fearnside (2000) calculated a total net emission of carbon from tropical land uses, equivalent to approximately 29% of the total anthropogenic emission from fossil fuels and land-use change. These numbers emphasize the potential for C management if only biomass is utilized being ablaze each year.

As a result of intensive research done in the 1980s, carbonized materials are formally authorized for use as soil amendment material in Japan. Thus Japan used 27% of the total charcoal consumption (50,835 Mg in 1999) for purposes other than fuel. By far the biggest proportion was used for agriculture (30.6%) followed by livestock industries (22.3%), gardening and golf courses (7.6 %) (Okimori, et al., 2003). A Japanese company established charcoal production at an Indonesian tree plantation for pulp production. Their feasibility study with conventional charcoal-making methods showed that 77,000 Mg charcoal could be produced per year, and the carbon emission reductions by the project reaches 62,000 Mg C yr^{-1} (= 230,000 Mg CO_2 yr^{-1}) at an annual wood harvest of 10,000 ha (Okimori, et al., 2003).

Soil Fertility Enhancement

Antal and Grønli (2003) mentioned that most potting soils, herbicides in carbon-based formulations, and culture media formulations contain charcoal or activated carbons, although the scientific rationale for these applications is absent. Recent studies showed that soil charcoal amendments are indeed capable of increasing soil fertility. Charcoal significantly increased plant growth and nutrition in a pot experiment by Lehmann, et al. (2003) and a field experiment by Steiner, et al., (2007c). The authors proposed that charcoal can improve soil chemical, biological, and physical properties, but could not completely discern the mechanisms of fertility enhancement. Lehmann, et al., (2003) found significantly reduced leaching of applied fertilizer N in charcoal containing pots. This was corroborated by the findings of Steiner, et al., (2008a).

Soil respiration and the microbial population growth rate were found significantly altered by charcoal amendments (Steiner, et al., 2008b). Steiner, et al., (2004a) found increased microbial activity on charcoal amended plots. *Terra Preta* soils were marked by a very low soil respiration but very high population growth after substrate (glucose) additions. Unmanaged forest soils (Ferralsol) had a higher respiration rate but a very low population

[1] e-mail: Christoph.Steiner@biochar.org.

growth potential. These results reflect the relatively high biodegradable soil organic matter (SOM) content of primary forest topsoil but low available nutrients (requirement for microbial population growth), in contrast to refractory *Terra Preta* SOM with high available soil nutrient contents. Thus we conclude that nutrient availability in *Terra Preta* is independent from SOM decomposition. The effects on soil biology seem to be essential as charcoal has the potential to alter the microbial biomass (Steiner, et al., 2004a) and composition (Birk, 2005) and the microbes are able to change the charcoal's properties (Glaser, et al., 2001a). Rondon, et al. (2006) found increased biological N fixation by common beans through charcoal additions and Gehring (2003) increased occurrence of nodulating plants in forests on *Terra Preta* than on adjacent soils. The majority of experiments conducted show that charcoal soil amendments result in enhanced colonization rates my mycorrhizal fungi (Warnock, et al., 2007).

BIOCHAR PROSPECTS AND ESSENTIAL RESEARCH

The global potential of biochar (non fuel use charcoal) reaches far beyond *slash and char*. Inspired by the recreation of *Terra Preta*, most biochar research was restricted to the humid tropics. More information is needed on the agronomic potential of charcoal, the potential to use alternative biomass sources (crop residues) and production of by-products to evaluate the opportunities for adopting a biochar system on a global scale. Biochar as soil amendment needs to be studied in different climate and soil types. Today, crop residue biomass represents a considerable problem as well as new challenges and opportunities.

A system converting biomass into energy (hydrogen-rich gas) and producing charcoal as a by-product might offer an opportunity to address these problems. Charcoal can be produced by incomplete combustion from any biomass, and it is a by-product of the pyrolysis technology used for biofuel and ammonia production (Day, et al., 2005).

The acknowledgement of biochar as carbon sink would facilitate C-trading mechanisms. Although most scientists agree that the half life of charcoal is in the range of centuries or millennia, a better knowledge of the charcoal's durability in different ecosystems is important to achieve this goal. An access to the C trade market holds out the prospect to reduce or eliminate the deforestation of primary forest, because using intact primary forest would reduce the farmer's C credits. Fearnside (1997) estimated the above-ground biomass of unlogged forests to be 434 Mg ha^{-1}, about half of which is C. This C is lost if burned in a slash-and-burn scenario and lost at a high percentage if used for charcoal production. The C trade could provide an incentive to cease further deforestation; instead reforestation and recuperation of degraded land for fuel and food crops would gain magnitude. As tropical forests account for between 20 and 25% of the world terrestrial C reservoir (Bernoux, et al., 2001), this consequently reduces emissions from tropical forest conversion, which is estimated to contribute globally as much as 25 % of the net CO_2 emissions (Palm, et al., 2004).

Today most biomass gasification systems tend to suppress the creation of residuals, like total organic carbon (TOC) and ashes. C-emission trading options and a better knowledge of charcoal as soil additive would add value to these residues. Further, this would facilitate the use of alternative biomass, those which are currently avoided to due their higher TOC residuals. The tarry vapors constitute a significant loss of carbon during carbonization (Antal and Grønli, 2003), although representing another valuable product. Despite a lack of research, these condensed vapors are used for agricultural purposes mainly in Asia and Brazil (GERAIS, 1985; Glass, 2001; Steiner, et al., 2008b).

Japanese researchers attempt to produce charcoal with a specific pore size distribution to favor desired microorganisms (Okimori pers. communication). Pore structure, surface area, and adsorption properties are strongly influenced by the peak temperature during charcoal production (Antal and Grønli, 2003). Increasing porosity is achieved with increasing temperature but the functional groups are gradually lost. In this context, it is also important to discern the mechanisms of nutrient retention (mainly N) due to charcoal applications. The charcoal's low biodegradability (Kuhlbusch and Crutzen, 1995), low nutrient content (Ogawa, 1994; Antal and Grønli, 2003), and high porosity and specific surface area (Braida, et al., 2003) makes charcoal a rather exceptional SOM constituent. *Terra Preta* research has shown that oxidation on the edges of the aromatic backbone and adsorption of other organic matter to charcoal is responsible for the increased CEC, though the relative importance of these two processes remains unclear (Liang, et al., 2006).

Energy from crop residues could lower fossil energy consumption and CO_2-emissions, and become a completely new income source for farmers and rural regions. The biochar by-product of this process could serve to recycle nutrients, improve soils and sequester carbon. A review by Johannes Lehmann (2006) and the article "*Black is the new green*" (Marris, 2006) emphasize the potential of bio-char on a global scale. A global analysis by Lehmann, et al. (2006) revealed that up to 12% of the total anthropogenic C emissions by land use change (0.21 Pg C) can be off-set annually in soil, if slash and burn is replaced by slash and char. Agricultural and forestry wastes add a conservatively estimated 0.16 Pg C yr^{-1}. If the demand for renewable fuels by the year 2100 was met through pyrolysis, bio-char sequestration could exceed current emissions from fossil fuels (5.4 Pg C yr^{-1}). The described mixture of driving forces and technologies has the potential to use residual waste carbon-rich residues to reshape agriculture, balance carbon and address nutrient depletion.

References

Antal, M. J. and Grønli, M. *Ind. Eng. Chem. Res.* 2003, 42, 1619-1640.

Bernoux, M.; Graça, P. M. A.; Cerri, C. C.; Fearnside, P. M.; Feigl, B. J. and Piccolo, M. C. In *The biogeochemistry of the Amazon Basin*; McClain, M. E.; Victoria, R. L. and Richey, J. E.; Ed.; Oxford University Press: New York, NY, 2001; pp 165-184.

Birk, J. J. 2005, *Lehrstuhl für Bodenkunde und Bodengeographie*, University of Bayreuth, Diplomarbeit.

Braida, W. J.; Pignatello, J. J.; Lu, Y. F.; Ravikovitch, P. I.; Neimark, A. V. and Xing, B. S. *Environ. Sci. Technol.* 2003, 37, 409-417.

Day, D.;Evans, R. J.;Lee, J. W. and Reicosky, D. *Energy.* 2005, 30, 2558-2579.

Fearnside, P. M. *Climatic Change*. 1997, 35, 321-360.

Fearnside, P. M. *Climatic Change.* 2000, 46, 115-158.

Fearnside, P. M.; Lima, P. M.; Graça, A. and Rodrigues, F. J. A. *Forest Ecol. Management.* 2001, 146, 115-128.

Gehring, C. 2003, University of Bonn, *PhD theses.*

GERAIS, F. C. T. D. M. STI/CIT, Brasilia, *Documentos* 23, 1985, pp 18.

Glaser, B.; Balashov, E.; Haumaier, L.; Guggenberger, G. and Zech, *W. Org Geochem.* 2000, 31, 669-678.

Glaser, B.; Guggenberger, G.; Haumaier, L. and Zech, W. In *Sustainable management of soil organic matter*; Rees, R. M.; Ball, B. C.; Campbell, C. D. and Watson, C. A.; Ed.; CABI Publishing: Wallingford, UK, 2001a; pp 190-194.

Glaser, B.; Haumaier, L.; Guggenberger, G. and Zech, *W. Naturwissenschaften.* 2001b, 88, 37-41.

Glass, V. *Globo Rural.* 2001, vol. 188.

Goudriaan, J. In Carbon sequestration in the biosphere. Processes and prospects.; Beran, M. A.; Ed.; *Global environmental change*; Springer Verlag: Heidelberg, Berlin, 1995; 33, pp 3-18.

Katzer, F. *Grundzüge der Geologie des unteren Amazonasgebietes* (des Staates Pará in Brasilien). M. Weg: Leipzig, Germany, 1903.

Kuhlbusch, T. A. J. and Crutzen, P. *J. Global Biochem. Cy.* 1995, 9, 491-501.

Lehmann, J.; da Silva Jr., J. P.; Steiner, C.; Nehls, T.; Zech, W. and Glaser, B. *Plant Soil.* 2003, 249, 343-357.

Lehmann, J.; Gaunt, J. and Rondon, M. *Mitigation and Adaptation Strategies for Global Change*. 2006, 11, 403-427.

Liang, B.; Lehmann, J.; Solomon, D.; Grossman, J.; O'Neill, B.; Skjemstad, J. O.; Thies, J.; Luizão, F. J.; Petersen, J. and Neves, E. G. *Soil Sci. Soc. Am.. J.* 2006, 70, 1719-1730.

Lima, H. N.; Schaefer, C. E. R.; Mello, J. W. V.; Gilkes, R. J. and Ker, J. C. *Geoderma.* 2002, 110, 1-17.

Marris, E. *Nature*. 2006, 442, 624-626.

Ogawa, M. *Farming Japan*. 1994, 28-5, 10-30.

Okimori, Y.; Ogawa, M. and Takahashi, F. *Mitigation and Adaptation Strategies for Global Change*. 2003, 8, 261-280.

Palm, C.; Tomich, T.; Noordwijk, M. v.; Vosti, S.; Gockowski, J.; Alegre, J. and Verchot, L. *Environment, Development and Sustainability*. 2004, 6, 145-162.

Rondon, M. A.; Lehmann, J.; Ramírez, J. and Hurtado, M. *Biol Fertil Soils*. 2006, DOI 10.1007/s00374-006-0152-z.

Smith, H. H. Brazil, *the Amazons and the coast*; C. Scribner's Sons: New York, NY, 1879.

Sombroek, W. G. 1966, Wageningen, The Netherlands, PhD theses.

Steiner, C. Slash and Char as Alternative to Slash and Burn; PhD theses, Cuvillier Verlag Göttingen, Germany.

Steiner, C.; Glaser, B.; Teixeira, W. G.; Lehmann, J.; Blum, W. E. H. and Zech, W. J *Plant Nurtr. Soil. Sc.* 2008a, 171-6, 893-899.

Steiner, C.; Das, K.; Garcia M.; Förster B.; Zech, W. *Pedobiologia*. 2008b, 51, 359-366.

Steiner, C.; Teixeira, W. G.; Lehmann, J.; Nehls, T.; Macêdo, J. L. V. d.; Blum, W. E. H. and Zech, *W. Plant Soil.* 2007c, 291, 275-290.

Steiner, C.; Teixeira, W. G.; Lehmann, J. and Zech, W. In *Amazonian Dark Earths: Explorations in space and time*; Glaser,. B. and Woods, W. I.; Ed.; Springer Verlag: Heidelberg, 2004a; pp 195-212.

Steiner, C.; Teixeira, W. G. and Zech, W. In *Amazonian Dark Earths: Explorations in space and time*; Glaser, B. and Woods, W. I.; Ed.; Springer Verlag: Heidelberg, 2004b; pp 183-193.

Tiessen, H.; Cuevas, E. and Chacon, P. *Nature.* 1994, 371, 783-785.

Warnock D. D.; Lehmann J.; Kuyper T. W.: Rillig M. C.: Plant Soil. 2007, 300, 9-20.

Zech, W.; Haumaier, L. and Hempfling, R. In *Humic substances in soil and crop sciences; selected readings*; Ed.; ASA-SSSA: Madison, WI, 1990; pp 187-202.

In: Soil Fertility
Editors: D. P. Lucero and J. E. Boggs

SBN: 978-1-60741-466-7

Chapter 10

Fertility Status and Dynamics of Soils in the Nepal Himalaya: A Review and Analysis

***Roshan M. Bajracharya*[1] *and Dil Prasad Sherchan*[2]**
[1]Dept. of Environmental Science and Engineering,
Kathmandu University, Dhulikhel, Nepal
[2]Nepal Agricultural Research Council,
Kathmandu, Nepal

Abstract

The Himalayan mountain range stretches from Afghanistan in the west to Myanmar in the East covering a distance of over 3500 km. Nepal is a densely populated, highly diverse and ecologically fragile country situated within the Central Himalaya between India and China. The inhabitants of the Himalayan region depend heavily on subsistence agriculture as the mainstay and primary source of livelihood. Natural resources in Nepal, in particular, have come under tremendous pressures due to the ever-increasing human and livestock populations and demand for food, fiber, construction material and energy. This has led to degradation of the resource base including, forests, water and arable land. Numerous workers have reported a steady decline in the fertility status of agricultural and forest soils of Nepal. Moreover, high rates of soil erosion and nutrient loss have been reported from degraded forest land, grazing lands and sloping agricultural fields. Other factors contributing to the fertility and productivity decline of soils include: increasing trend of chemical fertilizer use, decreasing amounts and quality of farmyard manure/compost applied, and increasing cropping intensification. Sustainable options for maintaining and enhancing soil fertility and productivity must address increasing the return of crop residues and other organic materials to soil, reducing dependence on inorganic fertilizers, and shifting to alternative fertilizer sources (such as bio-fertilizers) and cropping systems (such as agroforestry and permaculture).

Keywords: physiographic regions, cropping system, land use, agro-ecology, nutrient loss, bio-fertilizer, crop productivity.

Introduction

Hill regions of the Himalaya are moderately densely populated areas with limited productive agricultural land and ever-increasing demands upon the land and land-based resources. The Himalaya stretch across Asia covering a length of over 3500 km from Afghanistan at its western extreme to Myanmar in the southeast and is home to over 150 million people (ICIMOD 2006). The main causes of environmental pressures in the region are the rising human and livestock populations, expanding semi-urban and urban areas leading to reduction in cultivable land area, and an increase in agricultural intensification to meet food and livelihood needs (ADB/ICIMOD 2006, Dahal et al. 2006, Shah and Schreier 1995).

The consequences of the increasing pressures upon the environment is steady degradation of the natural resource base, specifically the forests, pastures, soil, and water. The degradative processes are interrelated and reinforce one another usually initiated by the clearing and destruction of vegetation. Devoid of the protective and stabilizing vegetative cover, surface runoff of rain water increases while infiltration and percolation to the groundwater is reduced. Hence, soils become susceptible to the various forms of erosion, which removes fertile topsoil and concurrently reduces effective rooting depth (Lal 1997). Moreover, sediment and nutrient-laden stream flow may have adverse consequences downstream (Shah et al. 2000, Merz et al. 2000). Thus, the cycle of degradation, once set in motion, builds up momentum, and if left un-checked, may cross the critical threshold beyond which recovery of the fragile Himalayan ecosystems may be very difficult (Bajracharya et al. 2006, 2007).

Background

The Nepal Himalaya form the central part of the range stretching about 850 km from east to west with the Tibetan plateau to the north and Gangetic plains of India on the south (Figure 1). Agriculture remains Nepal's principal economic activity directly engaging around 66% of the population (23.3 million, 2001 census). The agricultural sector contributes to about 39% of the national GDP. None the less, the present food production falls short of the requirements of a growing population with annual growth rate of 2.25% (MOAC 2005). The Middle Mountains region is particularly vulnerable to food insecurity as it has a dense population (42% of total), limited prime agricultural land (only about 16% of the total area), fragile ecology, active geology, and a predominance of subsistence farming on marginal (steep) lands with poor quality soils.

Traditionally, farmers in the Nepal hills have relied upon compost or farmyard manure (FYM, made of forest litter, crop residues and animal manure) to replenish crop lands, and in the past, the less intensive 2-crop with fallow farming systems had been sustainable for generations. Over the past two or three decades, however, a shift in cropping patterns spurred on by increasing food and cash-crop demands along with easy availability of agro-chemicals, has led to diminishing productivity and fertility of arable lands (Bajracharya 2002, Regmi et al. 2005, Karki 2006). Intensified vegetable cropping systems with upto 4 crops grown per annual cropping cycle have become commonplace especially in peri- and semi-urban areas in close proximity to highways and urban markets. The production of major cereal crops have been shown to be virtually stagnant over the past 15 to 20 years (Figure 2a) and the national average yields of maize, wheat and rice are well below the attainable and experimental yields (Figure 1b). Indeed, the productivity of these crops are well below those reported in neighboring countries (Kaini 2004). The main reasons for low yields are believed to be lack of replenishment of soil organic matter and inadequate use of fertilizers (Sherchan and Karki 2005).

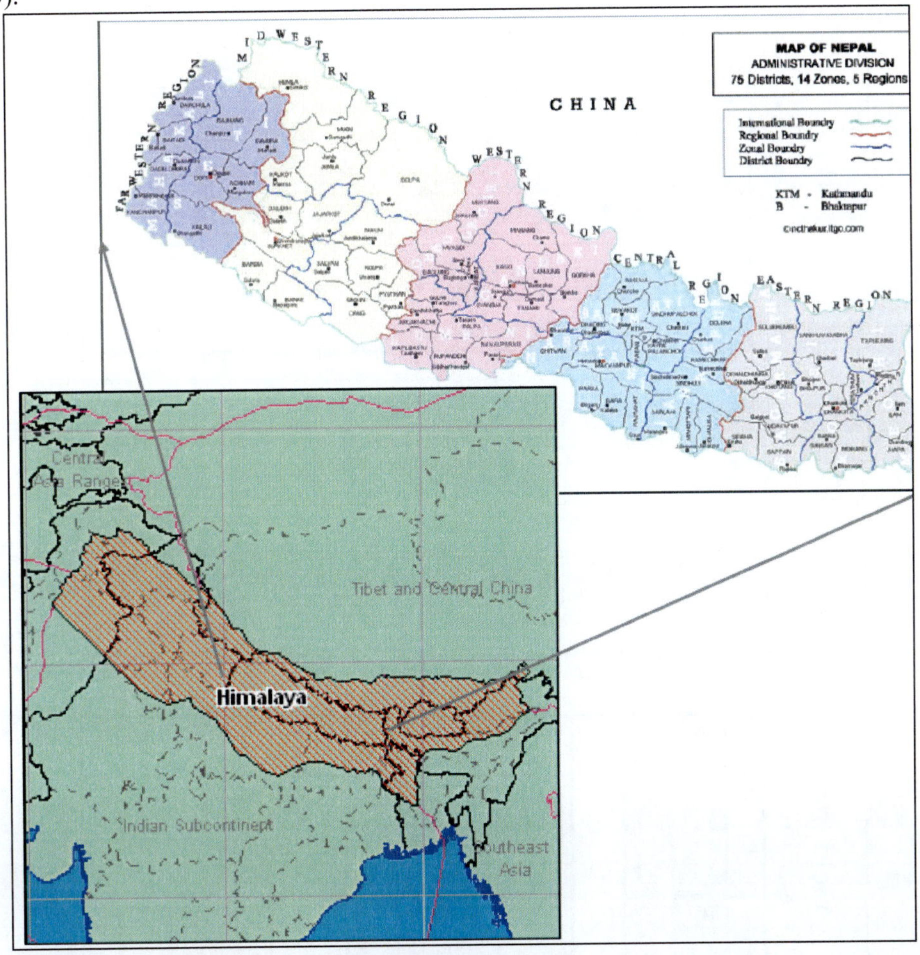

Figure 1. Map showing the Himalayan mountain range and administrative divisions of Nepal. (Source: www.peakbagger.com; http://ncthakur.itgo.com).

OVERALL FERTILITY STATUS OF SOILS IN NEPAL

A comprehensive compendium of the fertility status of soils covering all five development regions from east to west, as well as, hills and plains, was compiled at a workshop to assess soil fertility management in Nepal by the Department of Agriculture in collaboration with JICA (Japan International Cooperation Agency) in 1999 (DOA/STSS 1999). Although data for thousands of samples analyzed at the various Regional Agricultural Research Centers (RARCs) was summarized, only qualitative information regarding five chemical parameters, namely soil pH, organic matter (OM), total nitrogen (TN), available phosphorous (P_2O_5) and exchangeable potassium (K_2O) was provided (Table 1).

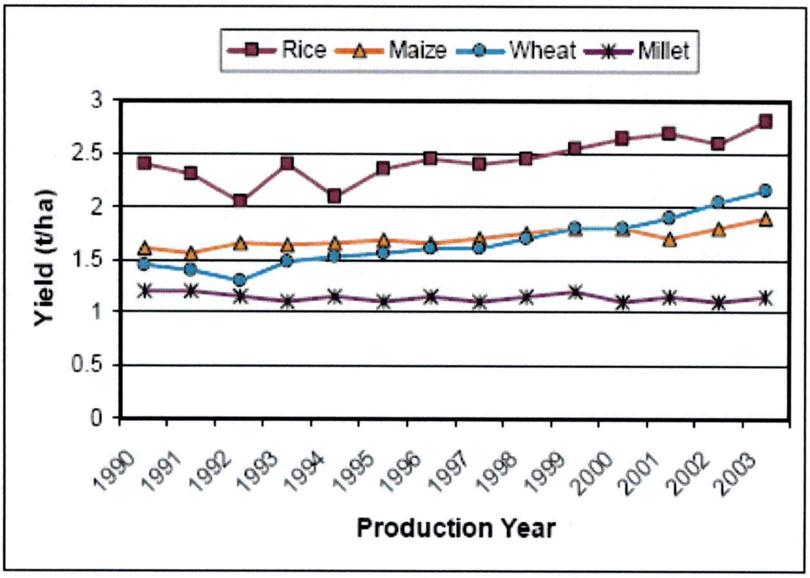

A

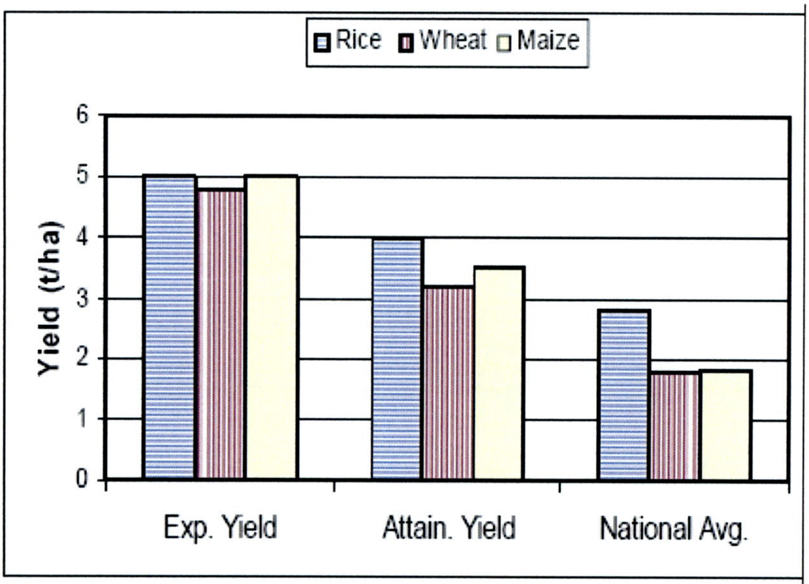

B

Figure 2. Productivity of major cereal crops in Nepal and the existing yield gap. (Source: redrawn from MOAC 2005).

On the basis of the information presented, it could be deduced that the majority (over 90%) of soils in Nepal fall into the low to medium classes with respect to OM and TN status. Also, a good proportion of soils (55%) are acidic in nature, while the majority tends not to be low in available P or exchangeable K (Table 1).

Table 1. Overall fertility status of soils in five development regions of Nepal (percent of soils in various classes)

Development	Soil pH			Organic Matter			Total Nitrogen			Available P			Exchangeable K		
Region§	Acidic	Neutral	Basic	Low	Medium	High	Low	Medium	High	Low	Medium	High	Low	Medium	High
Data range	<6	6-7.5	>7.5	<2%	2-4%	>4%	<0.1%	0.1-0.2	>0.2%	<50ppm	50-200	>200	<50ppm	50-200	>200
EDR - plains (n=913)	66	12	22	80	17	3	73	23	4	27	25	48	38	41	21
EDR - hills (n=194)	87	10	3	37	54	9	23	55	22	11	16	73	13	25	62
CDR - plains (n=255)	31	53	16	44	55	1	33	65	2	27	34	39	47	42	11
CDR - hills (n=748)	87	12	1	54	42	4	39	52	9	24	25	51	24	39	37
WDR (n=2100)	65	34	1	43	50	7	33	55	12	34	22	44	6	31	63
MWDR - hills (n=252)	31	28	31	80	17	3	79	18	3	50	24	26	62	18	20
MWDR - plains (n=980)	40	36	24	63	28	9	52	38	10	61	21	18	54	27	19
FWDR (n=1000)	37	35	28	78	21	1	63	33	4	50	22	28	4	50	46
Nepal means (n=6442)	56	28	16	60	36	5	49	42	8	36	24	41	31	34	35
Standard dev.	23.8	15.2	12.5	17.9	16.6	3.3	20.6	16.9	6.6	16.7	5.1	17.4	22.4	10.6	20.3

Source: DOA/STSS (1999), Tipathi (1999); n = number of soil samples.

§ EDR - Eastern development region, CDR - Central development region, WDR - Western development region, MWDR - Mid-western development region, FWDR - Far-western development region.

Another extensive study of over 9,000 soil samples from around the country augments the data presented in Table 1. On the basis of analyses conducted at the various RARCs, Regmi et al. (2005) categorized the general fertility status of soil as shown in Figure 3. This study confirms the above findings that a majority of Nepalese soils are of low to medium status in TN and soil OM. Only 5% of samples had high (>0.2%) TN and 11% of samples were high (>3%) in OM levels. On the other hand, soil available P and exchangeable K were observed to be mostly in the medium to high range, with only about a third of soils having low levels of these nutrients. Thus, based on the exhaustive records of soil samples analyzed by the Nepal Agricultural Research Council (NARC) to date, soils of many parts of the country are generally low in nitrogen and organic matter, which suggests on-going unsustainable farming practices and likely detrimental effects on crop productivity in the long run.

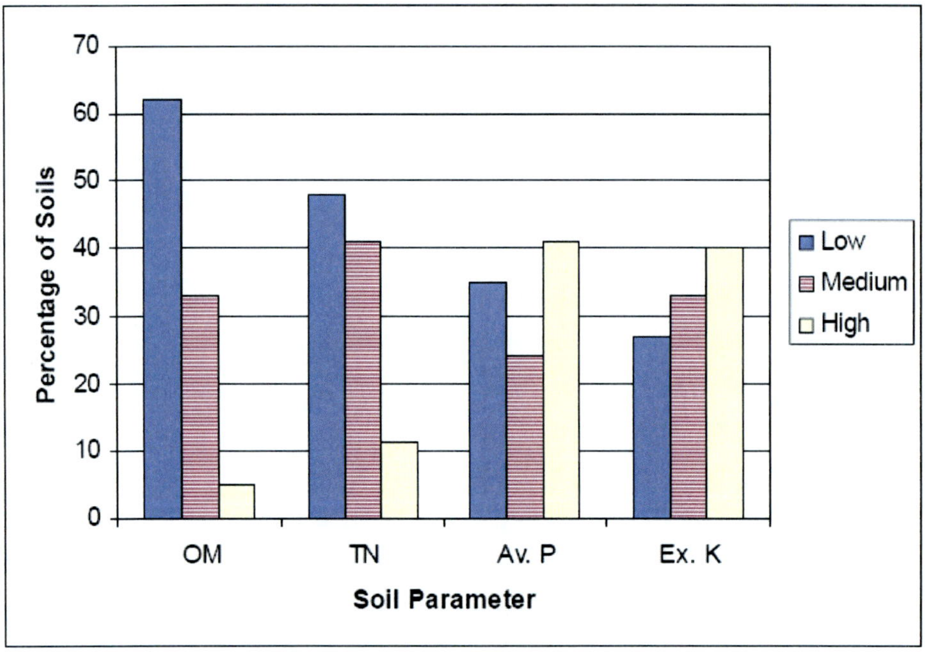

Figure 3. General status of soil organic matter (OM), total nitrogen (TN), available phosphorus (Av. P), and exchangeable potassium (Ex. K) in the Nepal hills. Note: n = 7520 for OM, 9872 for TN, 8942 for Av. P and 9522 for Ex. K. (Source: Regmi et al. 2005).

The levels of micronutrients (B, Cu, Fe, Mn, Mo, and Zn) in soil have been studied only to a limited extent in Nepal, but have been noted to be unmanaged for the most part (Andersen 2000). Tables 2 and 3 present the means and standard deviations and general status of micronutrients in soils from about 20 districts. With the exception of iron and manganese contents in soils, all other micronutrients were noted to be low in the samples analyzed (Karki et al. 2005). In particular, boron, copper, zinc and molybdenum were observed to be low in nearly all districts sampled (Tables 2 and 3).

Table 2. Micronutrient status of soils (ppm) from several districts in Eastern, Central and Western Nepal

	B	Cu	Fe	Mn	Zn
District	(mg/kg soil)				
Chitwan	0.30	2.41	70	11	0.86
Dhankuta		2.27	107	12	1.21
Gorkha	0.23	0.05	206	29	1.19
Kaski	0.61	1.94	188	67	0.83
Kathmandu	0.01	10.03	695	78	6.83
Myagdi	0.72	1.83	181	85	0.86
Palpa	0.93	1.18	225	36	0.72
Parbat	0.46	1.52	186	56	1.27
Tanahun	0.65	1.43	159	32	0.85
Mean	0.49	2.52	224	45	1.62
Std. Dev.	0.30	2.90	183	27	1.96

Note: B – boron, Cu - copper, Fe - iron, Mn - manganese, Zn – zinc.
Sources: Tripathi 2001, Karki et al. 2005, Sapkota & Andersen 2005.

Table 3. Plant available micronutrient status of soil in various districts of three physiographic

Regions of Nepal	B	Cu	Fe	Mn	Mo	Zn
Terai (n = 10)	Low	Low-med.	Low-med.	Low-high	Tr.-high	Low-med.
Middle Mountains (n = 10)	Low	Medium	Medium	Low-med.	Tr.-high	Low
High Mountains (n = 2)	Low	Low	Low	Low	Low	Low
Nepal (n = 22)	Low	Low-med.	Low-med.	Low-high	Tr.-high	Low-med.

Note: B - boron, Cu - copper, Fe - iron, Mn - manganese, Mo - molybdenum, Zn – zinc.
n = number of districts; Tr. = trace amount.
Sources: Karki et al. 2005, Rai et al. 2005, Sapkota & Andersen 2005.

Soil Fertility Variation by Elevation and Physiographic Regions

A number of researchers have attempted to evaluate the soil fertility status of soils in relation to altitudinal and physiographical region effects in the Eastern, Central and Western Region of Nepal (DOA/STSS 1999, Tripathi 2000, Tripathi and Sthrestha 2000, Tuladhar 1995). In general, soils of the hills region (Hill and Middle Mountains) of Nepal were more acidic in nature (40-87% of soils sampled) compared to those of the plains (Terai) region (31-66% of soils) as can be seen from the data in Table 1. With regard to soil OM contents, differences appear less distinct; however, the plains region had a slightly higher overall proportion of soils falling in the 'low' category than the hills. Likewise, the hills region had a slightly higher proportion of soils with 'medium' and 'high' OM contents compared to the plains. This was likely due to the trend of adding more FYM/compost in the hill regions where chemicals are less available and the slow decomposition rates at higher elevations.

Both the plains and hills regions had similar proportions of soils with 'medium' TN contents, although the hills had somewhat higher numbers of soil with 'low' TN than the plains. The nitrogen status of soils presumable reflects the availability and amounts of chemical fertilizer addition, which is typically more in the plains than in the hills. Available P and exchangeable K levels were predominantly in the 'high' and 'medium' categories for hills soils, while plains soils had mostly 'low' to 'medium' levels (Table 1). The status of phosphorus and potassium appear to be more influenced by geology and soil type than by other factors.

The micronutrient status of soils, though low for the most part, seemed to be somewhat higher in the Middle Mountains as compared to the Terai (plains) and the High Mountains regions (Table 3). The High Mountains had 'low' levels of all extractable micronutrients likely due to the slow weathering and soil formation process with consequently shallow and less developed soils in this region. The levels of boron were 'low' across all three physiographic regions, while iron and manganese were often 'medium' to 'high' in the Terai and Middle Mountains. However, within the Middle Mountains, elevation had no clear effect on extractable boron or zinc as can be inferred from Figure 4c.

In western Nepal, soil OM, TN and available P all showed an increasing trend with elevation (Figure 4 a and b). Organic matter, and correspondingly total nitrogen, tends to be higher at high elevations due to cooler, moister conditions and slower decomposition rates compared to lowlands. Soil pH and exchangeable K, on the other hand, did not follow any distinct pattern. Likewise, increases in OM, TN and available P contents with altitude from 300 to 2300 m were observed under rice and wheat cropping patterns by Tuladhar (1995). This trend did not, however, hold true for maize and barley crops in western Nepal (Figure 5).

Fertility as Influenced by Cropping Systems, Agro-Ecology and Land Use

The cropping pattern on arable land could be expected to influence soil fertility in conjunction with agro-ecology as determined by dominant landscape positions, soil types, and slope aspects on which particular crops are grown. Indeed, the dominant crops grown in a region tend to be a function of the elevation and agro-climatic zone. For instance, rice and wheat are typically grown mainly in the valleys and low lands, while maize, millet and barley are grown mostly on the sloping upland areas. Wheat, however, is grown both in irrigated lowland areas, as well as, on upland terraces to considerably high elevations (high mountain region). Numerous studies have examined the effects of cropping systems on soil fertility parameters in eastern, central and western Nepal (Sherchan et al. 1997, Sherchan and Gurung 1998, Tripathi 2000, Tripathi and Shrestha 2000, Tuladhar 1995).

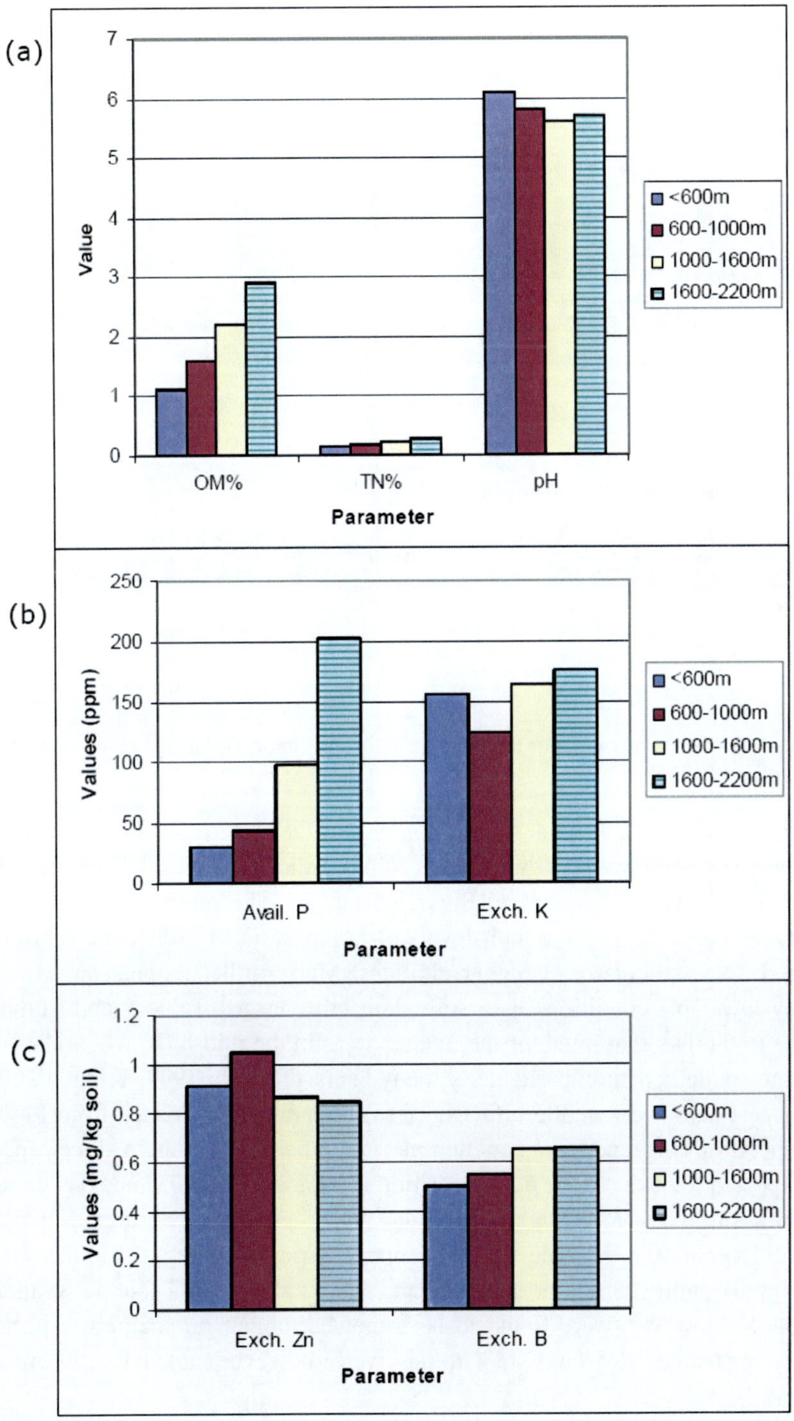

Figure 4. Variation of soil fertility parameters with elevation in western Nepal. (Source: Tripathi 2001).

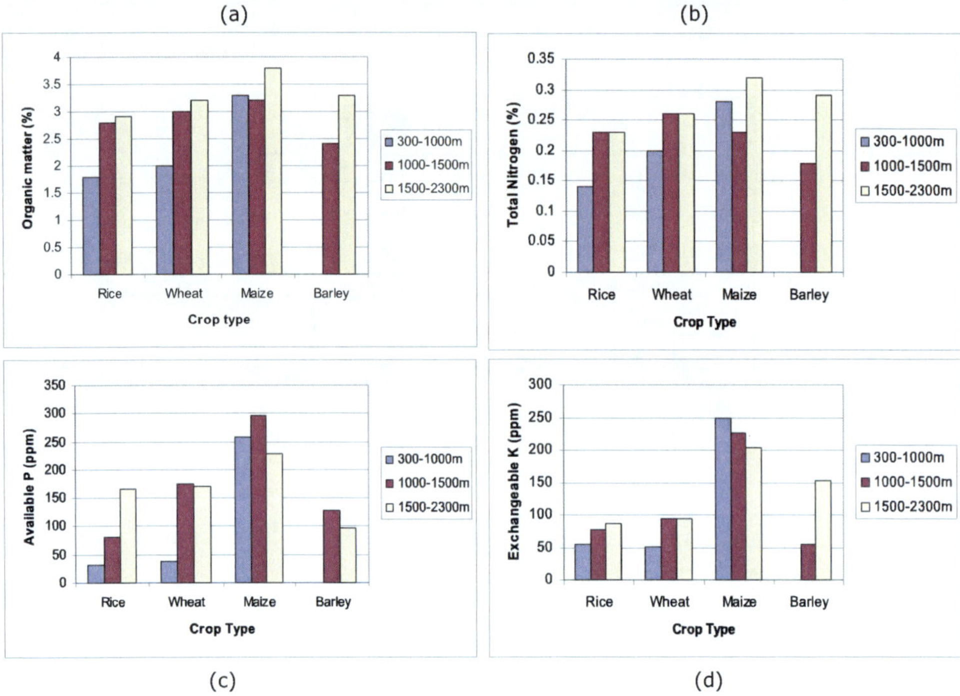

Figure 5. Influence of altitude on major plant nutrients in soils under dominant crops in western Nepal (Source: Tuladhar 1995).

All major nutrients N, P, and K, as well as soil OM, were found to be highest under maize cropping system in western and eastern Nepal, although OM and TN were only slightly less under rice and wheat than maize (Figures 5 and 6). The reason for this is that upland sloping terraces generally receive high levels of compost/FYM. Soil under barley also had high OM and TN, particularly at higher altitudes. Maize-millet cropped lands appeared to have slightly lower pH conditions than soils with other crops (Table 5 and Figure 6). This could be due to the fact that much of the upland areas in the mid-hills, where these crops are grown, tend to acidic in nature, particularly the red soils (Tripathi 1999, Tripathi 2000).

Slope aspect also had a subtle effect on fertility parameters as seen from Figure 7. The northerly aspects of hills tend to be cool and moist compared to southern slope aspects, hence would be expected to have slower decomposition rates and higher OM and TN contents. This was generally observed for rice, wheat and maize crops in western Nepal, although exceptions did occur, for example with soil under barley (Figure 7). Similar observations were made by Tripathi (2001) for soil OM and TN as shown in Table 4. Available P and exchangeable K, however, were found to be somewhat higher on southern aspects than for northward-facing slopes. No consistent trends were, however, noted for the micronutrient status of soils.

Examination of the soil fertility status by land use types based on data from several districts in the mid-hills (Table 5) indicates forest and shrub land soils to have lower pH and higher OM contents than cultivated soils. Expectantly, forest soils had the highest OM, and correspondingly, TN levels. However, a high variability observed in OM is due to forest condition (managed vs. degraded), stand density and regular collection of leaf litter and fodder by local farmers particularly from un-protected forests (Shah and Schreier 1995, Schreier et al. 2000).

Table 4. Effect of slope aspect on plant nutrients in soils of western Nepal hills

Slope aspect§	pH	OM	TN	Av. P	Ex. K	B	Cu	Fe	Mn	Zn
		--------- (%) ---------		---------- (mg/kg) ----------						
E	5.95	2.3	0.14	27	121	0.60	1.23	184	50	0.67
E/S	6.15	2.7	0.20	64	74	0.58	0.78	180	33	1.05
E/N	5.18	3.1	0.17	121	82	0.39	1.12	141	31	0.96
N	5.79	3.9	0.22	93	144	0.60	1.19	180	64	0.86
N/E	5.79	4.6	0.23	189	199	0.74	1.97	185	72	0.90
N/S	5.10	3.3	0.18	16	59	0.35	1.23	208	28	2.58
N/W	5.85	3.8	0.24	97	140	0.59	1.12	181	55	0.77
S	5.42	2.9	0.19	48	168	0.45	1.44	186	52	1.11
S/E	5.85	2.0	0.14	19	164	0.48	1.65	178	82	0.90
S/W	6.11	2.6	0.20	35	137	0.66	1.60	177	57	0.94
W	6.11	3.3	0.21	82	203	0.74	2.04	215	66	0.70
Mean	5.75	3.14	0.19	72	136	0.56	1.40	183	54	1.04
Std. dev.	0.37	0.76	0.03	52	48	0.13	0.38	19	17	0.53

§ E - east, S - south, N - north, W - west; OM - soil organic matter, TN - total soil nitrogen, Av. P – available, phosphorus, Ex. K - exchangeable potassium, B - boron, Cu - copper, Fe - iron, Mn - manganese, Zn – zinc. Source: Tripathi (2001).

Table 5. Soil fertility status and erosion rates according land use types in Nepal

Land Use	Soil pH	OM	TN	Av. P	Ex. K	Soil loss	Min. SL	Max. SL
		------- (%) -----		----- (mg/kg) -----		---------------- (t/ha) -------------		
Upland	5.8	2.6	0.15	159	417	28	2.0	105
	±0.6	±1.5	±0.08	±151	±281	±29		
Lowland	5.5	2.4	0.14	171	205	0.7	0.0	2.7
	±0.5	±1.0	±0.07	±154	±126	±0.9		
Forest	5.0	3.9	0.17	127	217	5.1	0.2	15.3
	±0.6	±1.9	±0.08	±102	±108	±4.7		
Shrub	5.3	3.5	0.16	144	191	58	0.4	420
	±0.8	±2.1	±0.08	±112	±124	±78		

Note: OM - soil organic matter, TN - total soil nitrogen, Av. P - available phosphorus, Ex. K - exchangeable potassium, SL - soil loss; ±standard deviations.

Source: Recalculated from numerous studies (Bajracharya et al. 2006, 2007, Dahal et al. 2006, Sherchan et al. 2003, Baral et al. 1999, Tripathi 1999, Howell et al. 1986).

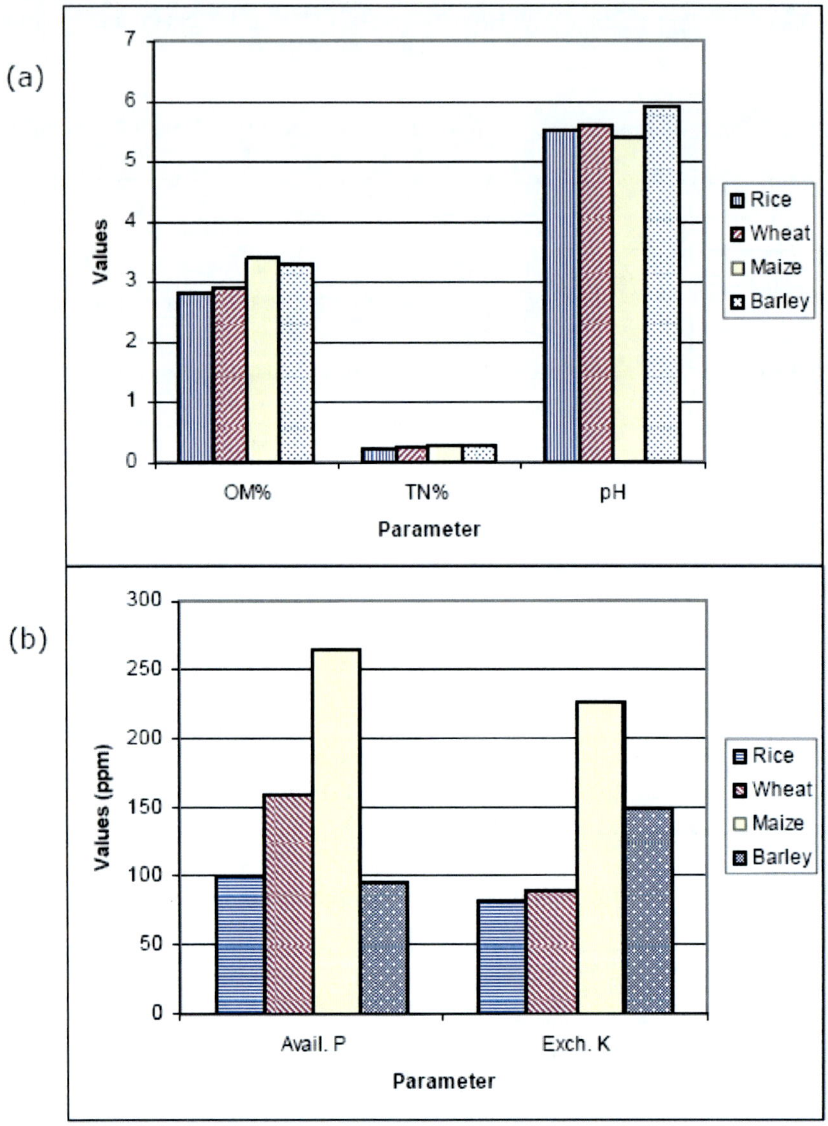

Figure 6. Fertility status of soils under different crop types in western Nepal. (Source: Tuladhar 1995).

Shrub land, unless degraded, also tends to have good vegetative cover and hence high levels of soil OM. Among the agricultural soils, upland areas had somewhat higher pH and OM, on the whole, than lowland (paddy-wheat) soils. This is thought to be due to the fact that lowland paddy-vegetable areas tend to receive more chemical fertilizer, mainly urea as basal and top-dressing, than upland rain-fed terraces, which could have contributed to the acidic conditions of lowland soils. The levels of total N, available P and exchangeable K did not differ greatly among the land uses and were highly variable as seen from the large standard deviations (Table 5).

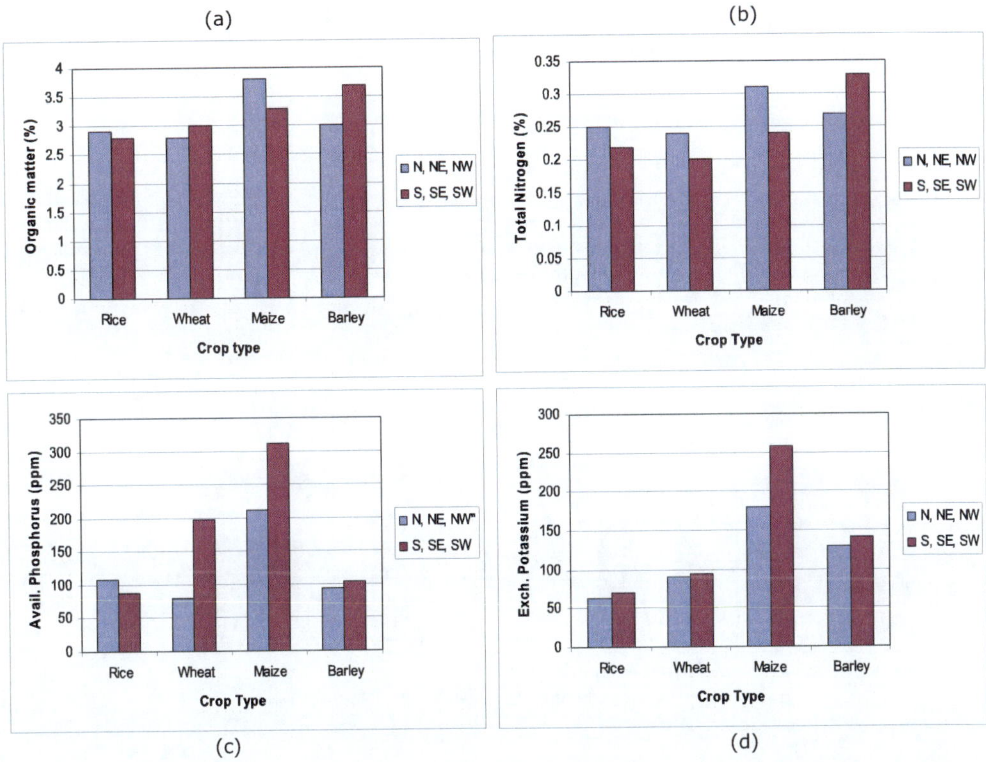

Figure 7. Effect of slope aspect on major soil fertility parameters in western Nepal. (Source: Tuladhar 1995).

Soil erosion rates for the hills and mountain regions of Nepal vary widely both spatially and temporally due to the diverse land use, soil, slope and climatic characteristics (Chalise and Khanal 1997). Differences in the surface runoff of rain water, and hence erosion, could be due either to management, soil types and land conditions or to scale of measurement, i.e., plot versus watershed scale (Chalise and Khanal 1987, Merz et al. 2000, Merz 2004). Annual soil loss rates from agricultural land ranged from 0 (lowland areas) to about 105 t/ha (sloping uplands) depending mainly on factors such as soil surface conditions and cover, slope and rainfall characteristics (Table 5). Lowland paddy areas have level bench terraces with minimal soil loss, while upland areas have sloping terraces of various widths and slopes, and hence, higher, more variable soil loss rates. Also, a distinct pattern of greatest proportion of the annual soil loss being concentrated in the early monsoon period (March-June) for agricultural land has been observed by numerous workers (Nakarmi et al. 2000, Carver and Nakarmi 1995, Atreya et al. 2004). This is because agricultural soils are left bare, freshly ploughed, and vulnerable at this time when pre-monsoon rains tend to be intense.

Soil loss from forest areas was least variable, with a comparatively narrow range of values from 0.3 to 15.3 t/ha/y (Table 5). The main determinants of runoff and soil erosion rates from forest land are tree density, crown cover, ground vegetation/litter cover and slope steepness. However, the coverage of ground vegetation and leaf litter appears to be most influential. In contrast the widest variation in soil loss rates occurred on grazing or pastures lands, ranging from 0.4 to 420 t/ha/y (Table 5). Such high rates of soil loss from degraded grazing lands could result from soil type (clayey lateritic), steep slopes, compaction of the surface layer, and exposure due to lack of vegetation, resulting in the formation of large gullies (Pahari 1993, Sah 1996, UNEP 2001).

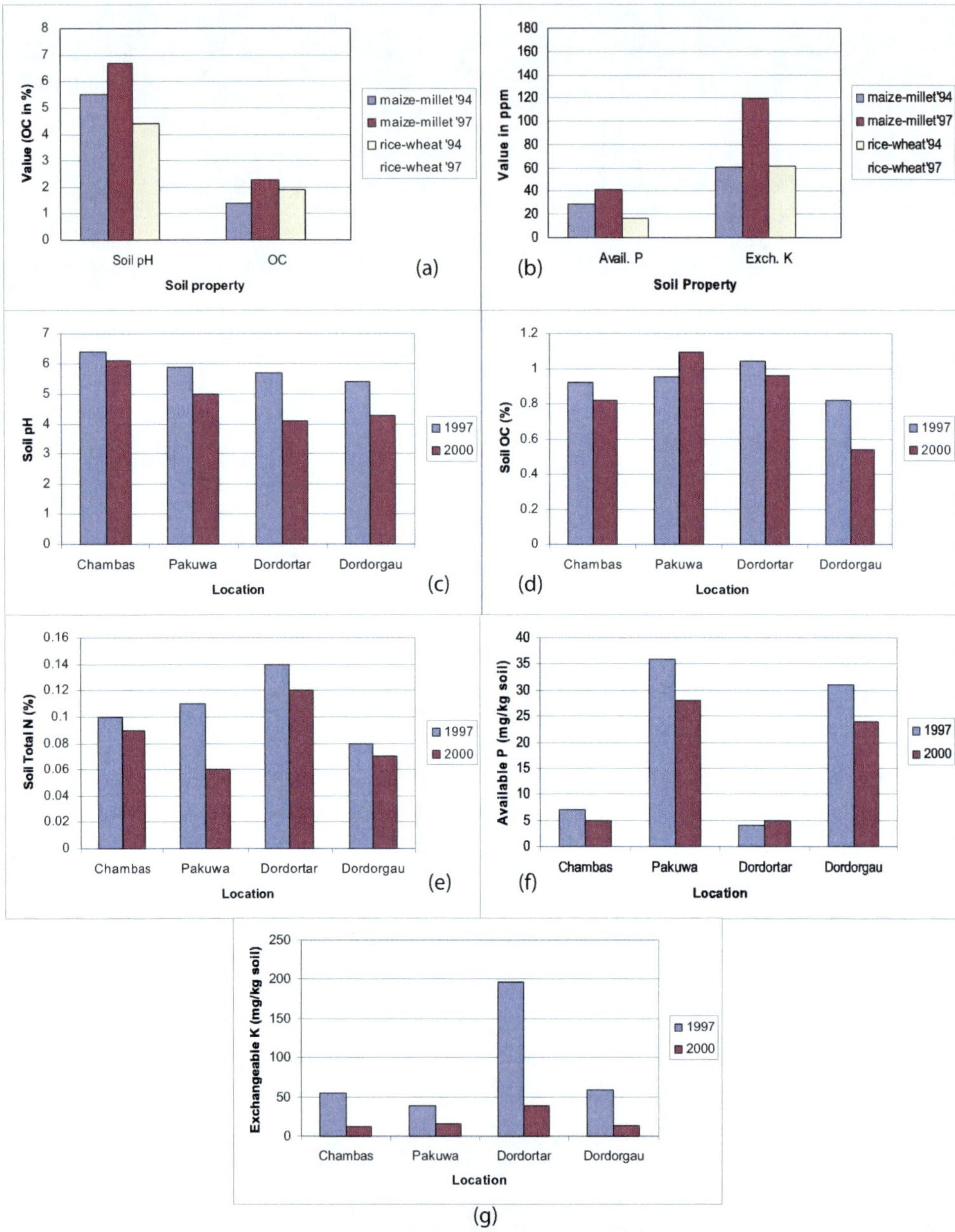

Figure 8. Short-term changes in soil fertility due to application of organic and inorganic fertilizers in eastern (a, b) and western (c – g) Nepal. (Sources: Sherchan et al. 1997, Sherchan & Gurung 1998, Tripathi & Shrestha 2000).

CHANGES IN FERTILITY STATUS OF SOILS IN NEPAL OVER TIME

In Nepal, chronological and detailed information on soil nutrient status is unavailable as systematic investigation of soil fertility is lacking. None the less, changes in the plant nutrient levels of soils may be analyzed by comparing the amounts of these nutrient elements in a benchmark study by FAO (Sillanpaa 1982) to those of more recent studies conducted after 1993. The FAO study provides an overview of the general status of both macronutrients and micronutrients in soils of Nepal prior to 1982 (Table 6). The data indicate that the soils were, on the whole, neutral in reaction; low in OM, available P, boron, molybdenum and zinc; medium in total N, exchangeable K, copper, and manganese; and, high in calcium, magnesium, and iron. Comparison with more recent soil fertility data (Tables 1, 2, 5 and Figure 3) suggest that the overall status has not changed greatly. The levels of soil OM and TN are similar, while soil pH appears to have decreased somewhat over the 20-25 year period (from greater than 6 to less than 6 on average). The improvement of soil OM and TN status of long-term fertility experiment at the Regional Agriculture Research Centers is thought to be due to better management and the use of integrated nutrient management (Table 7). Despite high variability, the P and K levels appear to have in fact increased appreciably, possibly due to increasing use of chemical fertilizers. The micronutrient levels appear also to remain unchanged.

Other workers have also studied short-term changes in soil fertility status and crop productivity, particularly in the Western and Eastern regions of Nepal (Tripathi et al. 2000, Tripathi and Shrestha 2000, Sherchan et al. 1997, Sherchan and Gurung 1998). In western Nepal, Tripathi and Shrestha (2000) assessed the change in soil fertility due to the application of organic and inorganic fertilizers.

Table 6. Benchmark study on status of soil nutrients in Nepal (from Sillanpaa, 1982)

Parameter	Mean	Std. dev.	Min.	Max.	n
Soil pH	6.16	0.91	4.5	7.69	50
Soil OM (%)	1.72	0.70	0.70	3.44	50
CEC (cmolc/kg)	19.7	5.7	9.5	36.5	50
Total N (%)	0.119	0.042	0.061	0.222	50
Avail. P (mg/l)	21.6	26.6	1.2	121.4	50
Exch. K (mg/l)	80	48	23	232	50
Calcium (mg/l)	1766	1173	220	4660	50
Magnesium (mg/l)	205	111	38	458	50
Boron (mg/l)	0.21	0.07	0.11	0.38	35
Copper (mg/l)	6.8	7.2	1.3	39.0	35
Iron (mg/l)	363	254	102	1150	35
Manganese (mg/l)	21.2	19.9	4.5	78.0	35
Molybdenum (mg/l)	0.063	0.066	0.012	0.320	35
Zinc (mg/l)	1.6	2.1	0.2	11.0	35
n = number of samples					

Table 7. Changes in nutrient status of soil in long-term fertility experiments in Nepal

Year	pH	OM(%)	TN	Av. P(kg/ha)	Ex. K	Data source
1978-79	6.8	1.3	0.075	122	166	Yadav et al., 1999 (Tarahara, Sunsari); rice-wheat-fallow
	±0.08	±0.28	±0.01	±33	±39	
1996-97	6.54	2.29	0.114	72	135	
	±0.11	±0.22	±0.007	±29	±23	
1978-79	8	1.03	0.09	9.8	124	Regmi et al., 2000 (Bhairahwa, Rupandehi); Rice-rice-wheat
	--	--	--	--	--	
1996-97	8.1	0.92	0.094	4.4	18.2	
	±0.07	±0.32	±0.032	±3.8	±5.2	
1978-79	6.87	--	0.049	156	417	Bhattarai and Mishra, 1999 (Khajura, Nepalgunj); rice-wheat
	±0.17	--	±0.01	±30	±35	
1996-97	6.98	1.6	0.057	35	115	
	±0.19	±0.32	±0.011	±22	±19	
1988-89	6.46	0.84	--	--	--	Gami and Sah, 1999 (Parwanipur, Birgunj); rice-wheat
	±0.18	±0.15	--	--	--	
1996-97	6.19	1.07	0.089	39	70	
	±0.24	±0.23	±0.026	±17	±8	

Note: OM - soil organic matter, TN - total soil nitrogen, Av. P - available phosphorus, Ex. K - exchangeable potassium; ±standard deviations.

Table 8. Changes in plant nutrients in soils of five districts of Nepal

Year	pH	OM(%)	TN	Av. P(kg/ha)	Ex. K	Data source
1993-94	5.46	3.11	0.151	122	172	Pandey et al., 1999 (Khumaltar, Kathmandu); rice-wheat
	±0.45	±0.35	±0.017	±23	±32	
1997-98	5.41	2.54	0.137	67	152	
	±0.31	±0.53	±0.016	±12	±11	
1993-94	6.02	2.31	0.096	22	788	Shakya et al., 1999 (Kabre, Dolakha); maize-millet
	±0.06	±0.32	±0.014	±3	±186	
1997-98	5.26	1.69	0.098	134	580	
	±0.08	±0.22	±0.01	±19	±69	
1993-94	5.3	1.62		20	174	Sherchan and Gurung, 1998 (Pakhribas, Dhankuta); Rice-wheat
1997-98	5.21	2.02		32	74	
	±0.08	±0.11		±3.8	±5.6	
1993-94	5.97	2.47	0.055	78	35	Tuladhar et al., 1999 (ARS, Sukhet); maize-wheat
	±0.14	±0.18	±0.007	±11	±9	
1997-98	5.9	1.44	0.096	37	55	
	±0.48	±0.24	±0.015	±3	±14	
2000	3.13	0.74	0.049	45	340	Atreya et al., 2006 (Dhulikhel, Kavre); maize-millet
	±0.21	±0.38	±0.02	±17	±138	
2002	3.31	0.7	0.079	92	228	
	±0.27	±0.21	±0.014	23	±63	

Note: OM - soil organic matter, TN - total soil nitrogen, Av. P - available phosphorus, Ex. K - exchangeable potassium; ±standard deviations.

They observed a generally decreasing trend in soil pH, organic carbon, and total nitrogen over a period of 4 years at four different locations as shown in Figure 8 c-g. In contrast, however, Sherchan et al. (1997) and Sherchan and Gurung (1998) noted an increasing trend in soil pH, OM, available P and K, and exchangeable Ca and Mg over a four-year period under both maize-millet and rice-wheat cropping systems in eastern Nepal (Figure 8 a and b). Available P and exchangeable K decreased in almost all cases (Table 8), presumably due to comparatively lower applications of P and K fertilizers as compared to N. The status of micro-nutrients, namely, iron, manganese, copper and zinc all, however, were reported to have declined in the same period (data not shown). These results may be due to addition of adequate FYM/compost, and nitrogen fertilizers on the one hand, and neglect of micro-nutrient elements on the other.

Alternatives for Sustainable Fertility and Productivity Enhancement

The above review of available data on soil fertility status and it change in the Nepal Himalaya revealed that the overall plant nutrient levels in a majority of soils examined are low to medium. Moreover, long-term fertilizer trial indicate that over about 20 years of continuous cropping of rice and wheat with chemical fertilizers, the soil nutrient status did not improve substantially and in most cases available P, exchangeable K and micronutrients (B, Mo, and Zn) decreased. In maize-based cropping patterns of the uplands, fertility generally declined during 3-5 year trials. In farmer's fields of western Nepal, soil fertility also declined during short-term trial between 1997 and 2000 (Tripathi and Shrestha 2000, Tripathi 2001). Furthermore, during the past 2-3 decades, crop yields have been reported to be declining (Maskey and Joshi 1991, FAO 1994, Vaidya et al. 1995, Brown et al. 1999) while food demands continue to increase, thus, leading to a food deficit. Farmer's response to changes in demands and market structures is intensification of cropping and a shift towards higher value vegetable/cash crops, which puts further stress on the land resource base (ADB/ICIMOD 2006, Dahal et al. 2006). It is, therefore, amply evident that effective intensive and sustainable approaches to enhanced crop production are urgently required. This in turn invariably calls for increasing plant nutrient levels in cultivated soils as well as enhancing the efficiency of nutrient use by the crops.

Increasing the levels of plant nutrients in soils may be achieved by addition of chemical fertilizers, organic manures or a combination of the two. As agrochemicals are comparatively inaccessible to a majority of Nepalese farmers, either due to lack of reliable markets and transport or due to economic reasons, organic and bio-fertilizers that are locally available must be exploited. Numerous studies in Nepal and other countries in the region demonstrate the benefits of integrated nutrient management (INS) or integrated plant nutrient systems (IPNS). Tripathi and Tuladhar (2000), Acharya et al. (1999) and Acharya et al. (2000) all reported significant improvements in soil fertility status and yields of rice, wheat and maize in western Nepal as a result of the combined application of organic manure and moderate amounts of inorganic fertilizers. The use of low quality organic fertilizer (semi-decomposed organic materials) in acidic sandy soils was found to be beneficial in enhancing the residual fertility in the soil to subsequent crops (Tripathi and Tuladhar 1997 and 2000).

Likewise, Sherchan et al. (1997), Sherchan and Gurung (1998) and Sherchan et al. (1999), all reported varying levels of benefits of the integrated approach of adding different compositions of organic manures and inorganic fertilizers. Yield increases and improvements in soil OM, TN and K were noted for both wheat-rice and maize/millet cropping patterns. Although residual effects were not seen in all cases, Sherchan and Gurung (1998) reported significant residual benefits to rice after wheat. On the whole, INM led to a 35% increase in maize/millet yields and 16% more yield for both crops in rice-wheat rotations (Sherchan et al. 2000).

Advantages of INM/IPNS for sustaining the soil resource and in enhancing yields have also been reported by researchers elsewhere in the region (Takkar and Jalali 2005). In India, Sharma and Sharma (2004) concluded that a combination of chemical fertilizers and FYM along with inclusion of a legume in the crop rotation (rice-potato-mungbean) was highly effective in building up soil organic carbon, nitrogen, phosphorus and potassium levels. Similar findings were reported by Linquist et al. (2007) in Laos where addition of crop residues along with commercial fertilizers enhanced nutrient balances in the soil, especially during years of unfavorable (fluctuating) soil moisture levels.

Another promising approach to enhancing sustainability of soil fertility and efficiency of crop nutrient uptake is the use of biological agents such as plants, algae and bacteria. Quyen and Sharma (2003) concluded that a combination of green manure (*Sesbania* sp.), blue green algae, and *Psuedomonas striata* inoculations, along with FYM application, gave the highest yield rice. The yield (5.2 t/ha) of organically grown rice was more than that obtained using high chemical fertilizer doses.

Apart from organic matter management and biological approaches, tillage also influences plant nutrient status through its impact on losses due to runoff, soil erosion or leaching. Atreya et al. 2004 and 2006, found reduced tillage systems to be effective in minimizing soil and nutrient losses while maintaining yields at par with the conventional tillage practice on upland maize-based cropping in central Nepal. Pilbeam et al. (2004), on the other hand, observed that in intensified agriculture, regardless of tillage, leaching losses could be minimized through basal application of manure early in the season followed by nitrogen fertilization later.

In Pakistan, minimum tillage, along with FYM amendments led to an increase in soil P and K along with OM and TN, while enhancing maize yields (Iqbal et al. 2007). Zhu et al. (2002) found that a conservation tillage system (ridge-furrow with seasonal no-till) led to restoration of a degraded hill soil in southern China, improving its physical, chemical and biological characteristics. In another study in China, Xu et al. (2007) observed that application of pig manure with reduced rates of commercial fertilizer caused significant increases in soil OC, TN, microbial biomass carbon and microbial biomass nitrogen on no-till plots.

Additional options with good potential in the Nepalese context are: the use of biogas as a rural energy source and the resultant biogas slurry as an effective fertilizer; and, the adoption of SALT (sloping land agricultural technology) as a holistic and integrated approach to land, crop and livestock management. As there are already over 100,000 biogas plant in operation all over Nepal (Karki 2004, ADB/ICIMOD 2006), it is proven to be an acceptable option for many rural communities. Biogas has the dual benefits of reducing emissions of methane and carbon dioxide to the atmosphere, while saving forests and mitigating climate change, as well as, the slurry serving as a fertilizer rich in plant nutrients. The SALT approach is, in many ways, compatible with practices of farmers in the mid-hills as it incorporates terracing, contour operations, fodder hedge-row planting, agroforestry and mixed cropping (Tacio 1993).

Future Outlook, Priorities and Needs

There is no doubt that demands on the soil resources will continue to increase and that further intensification of agriculture will be needed to meet food demands in the future (Schreier et al. 1994, Pandey et al. 1995). These pressures will certainly be tremendous in the Himalayan region due not only the huge populations but also to the delicate balance in which ecosystems of this region exist (Schreier et al. 2000, Bajracharya 2002). Adding to the uncertainty of future changes in the land's productive capacity is the ever-escalating phenomenon of global climate change. It is, therefore, with considerable innovativeness and perseverance that farming communities in the Himalaya will have to develop more efficient and sustainable approaches to enhanced agricultural production than at present. In this daunting challenge, scientists and farmer must join hands and work in a mutually reinforcing way find the optimum solutions. Successful collaborative efforts will indeed depend on the ability to build contextual knowledge systems enabling the comprehension of the complexities of natural environments, the farming systems, as well as the socio-cultural fabric of the many diverse locations across the Himalaya (Andersen 2005). It is only through such contextual knowledge, developed by merging scientific knowledge systems with local knowledge systems, and the appropriate incorporation of indigenous technical knowledge (ITK) that effective and self-perpetuating changes in nutrient management will be adopted by local farmers (Kandel and Wagley 1999). And to further complement the efforts of farmers and scientists, the government must formulate and implement appropriate policies, subsidies and incentives to ensure the widespread adoption and evolution of effective and sustainable soil/crop/land management approaches (ADB/ICIMOD 2006, Sherchan and Karki 2005).

Conclusions

Soils of the Nepal Himalaya are inherently of predominantly low to medium fertility status and often acidic. The situation has been further aggravated by the increasing pressures on the soil resource and intensification of agriculture during the past few decades. Over about 20 years of continuous cropping with chemical fertilizers, the soil nutrient status of cultivated soils have not improved and available P, exchangeable K and micronutrients (B, Mo, and Zn) generally decreased. Long and short term trials at the RARCs indicate integrated management of plant nutrients through the judicious application of organic and inorganic fertilizers could lead to sustained soil fertility and crop productivity. Other holistic approaches such as use of biofertilizers, SALT and conservation tillage practices could also lead to sustainable enhanced agricultural production. In light of the fact that pressures upon the land resource base will continue to increase in future, compounded by the effects of global climate change, farmers, scientists and policy makers must work together in a synergistic manner for enhanced and sustained food production and for the benefit of the present society as well as future generations.

REFERENCES

Acharya GP, Tripathi BP, Munankarmi RC, Sherchan DP, Gregory PJ and Pilbeam CJ (1999) Integrated soil fertility management on upland rice (Ghaiya)-blackgram and rice-wheat systems in the hills of Nepal. *Lumle Seminar Paper* No. 1999/6, Lumle Agric. Research Center, Kaski, Nepal, 13 pages.

Acharya GP, Tripathi BP, Shrestha SP and Gregory PJ (2000) Nutrients management in maize-fingermillet system in the hills of Nepal. *Lumle Seminar Paper* No. 2000/8, Lumle Agric. Research Center, Kaski, Nepal, 7 pages.

ADB/ICIMOD (2006) Environment Assessment of Nepal – Emerging Issues and Challenges. *Publ. Asian Development Bank and Int'l. Centre for Integrated Mountain Dev.*, Kathmandu, Nepal, 224 pages.

Andersen P (2000) Beyond local knowledge and institutional reach: micronutrient disorders in Hill agriculture. *Paper presented at a Regional Conference on Physical Mobility and Development in the Mountains*, March 15-17, 2000, Kathmandu, Nepal.

Andersen P (2005) Building contextual knowledge: the interface between local and scientific knowledge. IN: Andersen P, Tuladhar JK, Karki KB and Maskey SL (eds.) Micronutrients in South and Southeast Asia, *Procs. Int'l. Workshop* held 8-11 Sep. 2004. Kathmandu, Nepal. P. 87-95.

Atreya K, Sharma S, Bajracharya RM and Rajbhandari NP (2004) Conserving soil in agro-ecosystems of Nepalese Central Himalayas, *Procs. Int'l. Conference on the Great Himalayas: Climate, Health, Ecology, Management and Conservation.* Held Jan. 12-15, 2004, Kathmandu, Nepal. P. 137-142.

Atreya K, Sharma S, Bajracharya RM and Rajbhandari, N.P (2006) *Applications of reduced tillage in hills of central Nepal.* Soil and Tillage Research 88:16-29.

Bajracharya RM (2002) Fertility and productivity parameters for soil from five mid-hill districts of central Nepal. *Procs. Int'l. Seminar on Mountains, Royal Nepal Academy of Science and Technology*, Kathmandu, Nepal. P. 141-150.

Bajracharya RM, Sharma S, Sitaula BK, Dahal BM, Rokaya K and Jeng A (2006) Assessment of soil quality using physiochemical and biological indicators in a mid-hill watershed of Nepal, *Procs. Int'l. Seminar on Environmental and Social Impacts of Agricultural Intensification in Himalayan Watersheds*. Held Oct. 15-17, 2006, Kathmandu, Nepal. P. 105-114.

Bajracharya RM, Sitaula BK, Sharma S and Jeng A (2007) Soil quality in the Nepalese context – an analytical review. *Int'l. Journ. Ecology and Envir. Sci.* 33(2-3):143-158.

Baral SR, Malla MB, Howell J (1999) *Reclaiming Rato Mato: A Study of the Rehabilitation of Red Clay Soils in the Nepal Middle Hills.* Dept. of Roads, HMG/Min. of Works and Transport, and Ecological Society, Kathmandu, 44 pages.

Bhattarai EM and Mishra R (1999) Effect of long term application of chemical fertilizer and manure on crop productivity and soil fertility under rice-wheat cropping pattern at Khajura, Nepalgunj IN: Maskey SL, Tripathi BP, Regmi AP, Tuladhar JK and Adhikary B (eds.) *Procs. Workshop on Long-term Soil Fertility Experiments*, held 11-13 Aug., Lalitpur, Nepal. P. 158-168.

Brown S, Schreier H, Shah PB and Lavkulich LM (1999) Soil nutrient budget modeling: An assessment of agricultural sustainability in Nepal. *Soil Use and Mgmt.*, 15:1-15.

Carver M and Nakarmi G (1995) The effect of surface conditions on soil erosion and stream suspended sediments. IN: Schreier H Shah PB and Brown S (eds.) Challenges in Mountain Resource Management, Processes, Trends and Dynamics in Middle Mountain Watersheds, Procs. Workshop, 10-12 April 1995, *Int'l. Center for Integrated Mountain Dev.*, Kathmandu, Nepal, P. 155-162.

Chalise SR and Khanal NR (1997) Erosion processes and their implications in sustainable management of watersheds in Nepal Himalayas. IN: *FRIEND '97– Regional Hydrology: Concepts and Models for Sustainable Water Resource Management.* Proc. Postojna, Slovenia Conference, Sept—October 1997. Int'l. Assoc. of Hydrologic Sciences, Publ. No. 246, P. 325-334.

Dahal BM, Sitaula BK, Bajracharya RM, Atreya K and Jeng A (2006) Soil and nutrient losses from intensified agricultural land in Ansikhola watershed in Kavre Distric of Nepal. *Procs. Int'l. Seminar on Environmental and Social Impacts of Agricultural Intensification in Himalayan Watersheds*. Held Oct. 15-17, 2006, Kathmandu, Nepal. P. 120-128.

DOA/STSS 1999. *Workshop on Soil Fertility Management Activities in Nepal – Past, Present and Future*. 11 June 1999, Department of Agric./Soil Testing and Service Section and Japan Int'l. Cooperation Agency. Kathmandu, Nepal, 18 pages.

FAO 1994. The collection and analysis of land degradation data in Nepal. Report of the Expert Consultation of the Asian Network on Problem Soils. Food and Agric. Organization/United Nations publication 1994/4, Bangkok, Thailand.

Gami SK and Sah MP (1999) Long-term soil fertility experiment under rice-wheat cropping system. IN: Maskey SL, Tripathi BP, Regmi AP, Tuladhar JK and Adhikary B (eds*.) Procs. Workshop on Long-term Soil Fertility Experiments*, held 11-13 Aug., Lalitpur, Nepal. P 12-34.

Howell JH (1986) Soils of the Forestry Research Project Trial Sites: Chautara, Panchkhal, Thulo Sirubari and Kadambas. FRP trail site report No. 8. *Soil Section, Forestry Research and Information Centre*, Babar Mahal, Kathmandu, Nepal.

ICIMOD 2006. Annual Report 2005. *Int'l. Center for Integrated Mountain Development*, Kathmandu, Nepal. 92pp.

Iqbal M, Anwar-ul-Hassan and Lal R (2007) Nutrient content of maize and soil organic matter status under various tillage methods and farmyard manure levels. *Acta Agriculturae Scandinavica*, 57:349-356.

Kaini BR (2004) IncreasingCrops Production in Nepal. In: Sherchan DP, Adhikari K, Batsa BK and Sharma D (eds.) *Procs 24th National Summer Crops Research Workshop on Maize Research and Production in Nepal*, June 28 to 30, 2004, NMRP and NARC, Kathmandu, Nepal. Pages 15-19.

Kandel DD and Wagley MP (1999) Some salient indigenous technology practices for watershed management in Nepal. Dep. Soil Conser. Watershed Mgmt., Nepal, Food and Agric. Org., UN. Publ. ASIAN WATMANET-PWMTA, Kathmandu, Nepal, 37 pages.

Karki KB (2004) Improvement of animal dung, biogas, environment and soil fertility. *Procs. Fifth Nat'l. Animal Sci. Conven.* Livestock Development for Socio-economic Empowerment. NASA, Kathmandu, Nepal, pages 241-246.

Karki KB (2006) Impact of cropping intensification on nutritional balance in Nepalese soils. *Procs. Int'l. Seminar on Environmental and Social Impacts of Agricultural Intensification in Himalayan Watersheds*. Held Oct. 15-17, 2006, Kathmandu, Nepal. P 27-33.

Karki KB, Tuladhar JK, Uprety R and Maskey SL (2005) Distribution of micronutrients available to plants in different ecological regions of Nepal. IN: Andersen P, Tuladhar JK, Karki KB and Maskey SL (eds.) Micronutrients in South and Southeast Asia, *Procs. Int'l. Workshop* held 8-11 Sep. 2004. Kathmandu, Nepal. P 17-29.

Linquist BA, Phengsouvanna V and Sengxue P (2007) Benefits of organic residues and chemical fertilizer to productivity of rain-fed lowland rice and to soil nutrient balances. *Nutr. Cycl. Agroecosyst.* 79:59-72.

Lal R (1997) Degradation and resilience of soils. *Phil. Trans. R. Soc.* Lond. B, 352: 997-1010.

Maskey RB and Joshi D (1991) Soil and nutrient losses under different soil management practices in the middle mountains of central Nepal. IN: Shah PB, Schreier H, Brown S and Riley KW (eds.) Soil Fertility and Erosion Issues in the Middle Mountains of Nepal. Proceedings of the Workshop held during April 22-25, 1991. *Int'l. Center for Integrated Mountain Dev.*, Kathmandu, Nepal, P. 105-120.

Merz J (2004) Water balances, floods and sediment transport in the Hindu Kush-Himalayas. Data analyses, modeling and comparison of selected meso-scale watersheds. Ph.D. Dissertation, University of Berne, Publ. Int'l. Center for Integrated Mountain Dev., Kathmandu, Nepal, 339 pages.

Merz J, Dongol BS, Weingartner R and Nakarmi G (2000) Impact of land use on generation of high flows in the Yarsha Khola watershed, Nepal. IN: Allen R, Schreier H, Brown S and Shah PB (eds.) *The People and Resource Dynamics Project: The First Three Years (1996-1999).* Int'l. Center for Integrated Mountain Dev., Kathmandu, Nepal. P. 185-198.

MOAC (2005) Statistical Information on Nepalese Agriculture (Time Series Information), Ministry of Agriculture and Co-operatives, Agri-Business Promotion and Statistics Division, Agri-Statistics Section, Singhdurbar, Kathmandu, Nepal.

Nakarmi G, Schreier H, Merz J and Mahatma P (2000) Erosion dynamics in the Jhikhu and Yarsha Khola watersheds in Nepal. IN: Allen R, Schreier H, Brown S and Shah PB (eds.). *The People and Resource Dynamics Project: The First Three Years (1996-1999).* Int'l. Center for Integrated Mountain Dev., Kathmandu, Nepal. P. 209-218.

Pahari KJ (1993) Soil Erosion Susceptibility Using Remote Sensing and GIS: *A Case Study of Andhikhola Watershed Nepal.* Asian Institute of Technology, Master Thesis No. NR 93-16, Bangkok Thailand.

Pandey PR, Pandey SP and Khadha YG (1999) Long-term effect of organic and inorganic fertilizer on rice-wheat system in rainfed lowland conditions of Khumaltar. IN Maskey SL, Tripathi BP, Regmi AP, Tuladhar JK and Adhikary B (eds.) *Procs. Workshop on Long-term Soil Fertility Experiments*, held 11-13 Aug., Lalitpur, Nepal. P 116-134.

Pandey SP, Tamang DB and Baidya SN (1995) Soil fertility management and agricultural production issues with reference to middle mountain regions of Nepal. IN: : Scheier H, Shah PB and Brown S (eds.) *Challenges in mountain resource management in Nepal: Processes, trends and dynamics in middle mountain watersheds.* Proc. workshop of the workshop, April 10-12, 1995, Int'l. Center for Integrated Mountain Dev., Kathmandu, Nepal, P. 41-49.

Pilbeam CJ, Gregory PJ and Sherchan DP (1999) Quantifying the yield benefit of integrated nutrient management from long-term wheat-rice and maize/millet rotations in the hills of Nepal. IN: Maskey SL, Tripathi BP, Regmi AP, Tuladhar JK and Adhikary B (eds.) *Procs.*

Workshop on Long-term Soil Fertility Experiments, held 11-13 Aug., Lalitpur, Nepal. P. 198-204.

Pilbeam CJ, Gregory PJ, Munankarmy RC and Tripathi BP (2004) Leaching of nitrate from cropped rainfed terraces in the mid-hills of Nepal. *Nutrient Cycling in Agroecosystems* 69:221-232.

Quyen NV and Sharma SN (2003) Relative effect of organic and conventional farming on growth, yield and grain quality of scented rice and soil fertility. *Archives of Agronomy and Soil Science*, 49:623-629.

Rai SK, Pandey SP, Khadka YG, Karki KB and Uprety R (2005) Mapping spatial zinc distribution in Rupandehi District, Nepal. IN: Andersen P, Tuladhar JK, Karki KB and Maskey SL (eds.) *Micronutrients in South and Southeast Asia*, Procs. Int'l. Workshop held 8-11 Sep. 2004. Kathmandu, Nepal. P 49-55.

Regmi AP, Pandey SP and Joshy D (2000) Effects of long-term application of fertilizers and manure on soil fertility and crop yields in rice-rice-wheat cropping system in Nepal, IN: Abrol IP, Bronson KF, Duxbury JM and Gupta RK (eds.) *Long-term Soil Fertilizer Experiments in Rice-Wheat Cropping Systems.* Rice-Wheat Consortium for the Indo-Gangetic Plains Paper Series 6. New Delhi, India.

Regmi BD, Poudel C, Tripathi BP, Schulz S and Dhital BK (2005) Managing soil fertility problems of marginal agricultural lands through integrated plant nutrient systems: Experiences from the hills of Nepal. IN: Andersen P, Tuladhar JK, Karki KB and Maskey SL (eds.) *Micronutrients in South and Southeast Asia*, Procs. Int'l. Workshop held 8-11 Sep. 2004. Kathmandu, Nepal. P 109-119

Sah BP (1996) Degradation and its scio economic impacts using RS and GIS: *A case study of Trijuga watershed Nepal,* M.Sc. Thesis, NR 96-20. Asian Institute of Technology, Bangkok, Thailand, 190 pages.

Sapkota K and Andersen P (2005) Commercial horticulture farming and its effects on soil fertility: A case study from peri-urban agriculture in the Kathmandu Valley. IN: Andersen P, Tuladhar JK, Karki KB and Maskey SL (eds.) *Micronutrients in South and Southeast Asia*, Procs. Int'l. Workshop held 8-11 Sep. 2004. Kathmandu, Nepal. P 153-165.

Schreier H and Shah PB (2000) Soil fertility status and dynamics in the Jhikhu and Yarsha Khola Watersheds. IN: Allen R, Schreier H, Brown S and Shah PB (eds.) *The People and Resource Dynamics Project: The First Three Years (1996-1999).* Int'l. Center for Integrated Mountain Dev., Kathmandu, Nepal. P. 281-289.

Schreier H, Shrestha B, Brown S and Shah PB (2000) Forest dynamics in Nepal: quality, quantity, and community forestry issues in middle mountain watersheds. IN: Allen R, Schreier H, Brown S and Shah PB (eds.) *The People and Resource Dynamics Project: The First Three Years (1996-1999).* Int'l. Center for Integrated Mountain Dev., Kathmandu, Nepal. P. 129-139.

Schreier H, Shah PB, Lavkulich LM and Brown S (1994) Maintaining soil fertility under increasing land use pressures in the middle mountains of Nepal. *Soil Use and Management*, 10:137-142.

Shah PB and Schreier H (1991) Nutrient deficiency and soil fertility issues. IN: Shah PB, Schreier H, Brown S and Riley KW (eds.) *Proceeding of the Workshop on Soil Fertility and Erosion Issues in the Middle Mountains of Nepal*, April 22-25, 1991. Integrated Survey Section, Topographical Survey Board, Kathmandu, Nepal and International Development Research Centre, Ottawa, Canada, P. 260-266.

Shah PB and Schreier H (1995) Maintaining soil fertility in agriculture and forestry. IN: Scheier H, Shah PB and Brown S (eds.) *Challenges in mountain resource management in Nepal: Processes, trends and dynamics in middle mountain watersheds.* Proc. workshop of the workshop, April 10-12, 1995, Int'l. Center for Integrated Mountain Dev., Kathmandu, Nepal. P. 171-182.

Shah PB, Schreier HE and Nakarmi G (2000) Rehabilitation of degraded lands. IN: Allen R, Schreier H, Brown S and Shah PB (eds.) *The People and Resource Dynamics Project, The First Three Years (1996-1999).* Int'l. Center for Integrated Mountain Dev., Kathmandu, Nepal, P. 139-147.

Shakya PR, Gharti DB and Upadhyay SR (1999) Effects of long-term application of organic and inorganic fertilizers on maize and millet yields under a maize/millet cropping pattern: the results of 1993-1997 seasons. IN: Maskey SL, Tripathi BP, Regmi AP, Tuladhar JK and Adhikary B (eds.) *Procs. Workshop on Long-term Soil Fertility Experiments*, held 11-13 Aug., Lalitpur, Nepal. P 143-157.

Sharma SN and Sharma SK (2004) Role of crop diversification and integrated nutrient management in resilience of soil fertility under rice-wheat cropping system. *Archives of Agronomy and Soil Science,* 50:345-352.

Sherchan DP, Gurung GB and Resham KC (1997) Effects of long term application of chemical fertilizer and manure on soil properties and crop yields under maize-millet cropping pattern: the results of 1996/97 season. PAC Working Paper No. 193, Pakhribas Agric. Center, Dhankuta, Nepal, 6 pages.

Sherchan DP and Gurung GB (1998) Effect of long term application of chemical fertilizers and organic manures on soil fertility and crop yields under rice-wheat system in the hills. *Nepal Ag. Res. Journal* 2:35-41.

Sherchan DP and Karki KB (2005) Pland nutrient management for improving crop productivity in Nepal. IN: *Improving Plant Nutrient Management for Better Farmer Livelihood, Food Security and Environmental Sustainability*, Procs. Regional Workshop, Beijing, China 12-16, December pages 41-57.

Sherchan DP, Pilbeam CJ and Gregory PJ (1999) Response of wheat-rice and maize/millet systems to fertilizer and manure applications in the mid-hills of Nepal. *Expl. Agric.* 35:1-13.

Sherchan DP, Bajracharya RM, Tiwari KR and Balla MK (2003) Soil survey report of Pokhare Khola sub-watershed, Dhading, Nepal. Himalayan Degradation Project, Institute of Forestry, Tribhuvan Univ./Agic. Univ. of Norway, Pokhara, Nepal, 52 pages.

Sillanpaa M (1982) Micronutrients and the nutrient status of soils: a global study. FAO Soils Bulletin No. 48. Food and Agriculture Organization of the UN, Rome, Italy. P. 242-250.

Tacio HD (1993) Sloping Agricultural Land Technology: a sustainable agroforestry shceme for the uplands. *Agroforestry Systems*, 22(2):145-152.

Takkar PN and Jalali VK (2005) Soil fertility problems and strategies to reduce them in the Himalayan region of India. IN: Andersen P, Tuladhar JK, Karki KB and Maskey SL (eds.) *Micronutrients in South and Southeast Asia*, Procs. Int'l. Workshop held 8-11 Sep. 2004. Kathmandu, Nepal. P. 133-142.

Tripathi BP (1999) *Review on acid soil and its management in Nepal.* Lumle Seminar Paper No. 99/1, Lumle Agric. Research Center, Kaski, Nepal, 11 pages.

Tripathi BP (2000) *Advances in soil fertility research in the hills of Nepal.* Lumle Seminar Paper No. 2000/19. Lumle Agric. Research Center, Kaski, Nepal, 17 pages.

Tripathi BP (2001) Soil fertility status in farmer's field of the western hills of Nepal. IN: Manadhar HK, Shrestha CL, Shrestha, RK and Pradhan, SM. (eds.) *Procs. 1st Convention of the Society for Agric.* Scientists, held March 29-31, 1999, Kathmandu, Nepal. P. 121-128.

Tripathi BP and Shrestha SP (2000) Nitrogen content in farmyard manure and its effects on the productivity and soil properties of rice-wheat, upland blackgram and maize-fingermillet systems. *Lumle Working Paper* No. 2000/14, Lumle Agric. Research Center, Kaski, Nepal, 8 pages.

Tripathi BP and Tuladhar JK (1997) *Effect of low and high quality organic material on maize yield and soil properties.* LARC seminar paper no. 97/13. Lumle Agric. Research Center, Kaski, Nepal, 9 pages.

Tripathi BP and Tuladhar JK (2000) Sustaining wheat productivity and maintaining soil fertility in maize-wheat system. Nepal Agric. Res. Journal, 4:31-36.

Tripathi BP, Acharya GP, Shrestha SP and Greogry PJ (2000) *Integrated nutrients management on rice in rice-wheat and upland rice-blackgram systems.* Lumle Seminar Paper No. 2000/18, Lumle Agric. Research Center, Kaski, Nepal, 8 pages.

Tuladhar JK (1995) Fertility status of farmer's field soil within the extension command area of Lumle Agricultural Research Centre. *LARC Working Paper* No. 95/47, Kaski, Nepal, 27 pages.

Tuladhar JK, Chaudhary RB and Chaudhary TN (1999) Long-term soil fertility experiment under maize-wheat cropping pattern in mid-western hill of Nepal. IN: Maskey SL, Tripathi BP, Regmi AP, Tuladhar JK and Adhikary B (eds.) *Procs. Workshop on Long-term Soil Fertility Experiments*, held 11-13 Aug., Lalitpur, Nepal. P 158-168.

UNEP (2001) Nepal: State of the Environment. UN Environment Programme in collaboration with HMG/Min. of Population and Environment, Int'l. Center for Integrated Mountain Dev. and United Nations Envir. Prog., Thailand, 211 pages.

Vaidya A, Turton C, Joshi KD and Tuladhar J.K (1995) A system analysis of soil fertility issues in the hills of Nepal: Implications for future research. IN: Scheier H, Shah PB and Brown S (eds.) *Challenges in mountain resource management in Nepal: Processes, trends and dynamics in middle mountain watersheds*. Proc. workshop of the workshop, April 10-12, 1995, Int'l. Center for Integrated Mountain Dev., Kathmandu, Nepal, P. 63-80.

Xu Y, Chen W and Shen Q (2007) Soil organic carbon and nitrogen pools impacted by long-term tillage and fertilization practices. *Comm. Soil Sci. and Plant Analysis*, 38:347-357.

Yadav CR, Bhujel RB, Prasai HK and Chaudhary AL (1999) *Long-term fertilizer trial on rice-wheat-fallow cropping system at Tarahara.* IN: Maskey SL, Tripathi BP, Regmi AP, Tuladhar JK and Adhikary B (eds.) Procs. Workshop on Long-term Soil Fertility Experiments, held 11-13 Aug., Lalitpur, Nepal. P 35-58.

Zhu B, Gao M, Liu G, Zhang J, Chen S and Zhang X (2002) Soil degradation and soil productivity restoration and maintenance in hilly land of southern China. *Arch. Acker-Pfl. Bodn.*, 48:311-318.

INDEX

A

B

C

D

E

F

G

H

I

J

K

L

M

N

Q

R

S

T

U

V

W

X

Y

Z